Handbook of Control Technologies for Hazardous Air Pollutants

Handbook of Control Technologies for Hazardous Air Pollutants

Robert Y. Purcell, Pacific Environmental Services, Inc.
Gunseli Sagun Shareef, Radian Corporation

Edited and adapted by JACA Corporation
for the Environmental Protection Agency

SCIENCE INFORMATION RESOURCE CENTER

HEMISPHERE PUBLISHING CORPORATION
Cambridge New York Philadelphia San Francisco Washington
London Mexico City São Paulo Singapore Sydney

This permanent edition includes the complete text of the Environmental Protection Agency Technology Transfer Handbook: *Control Technologies for Hazardous Air Pollutants* prepared under the direction of the Air and Energy Engineering Research Laboratory, Research Triangle Park, North Carolina. Under their direction a task force of eleven scientists and reviewers created the extensive text and graphic and tabular matter which was then adapted and condensed by JACA Corporation, Fort Washington, Pennsylvania.

Science Information Resource Center is a cooperative venture of J.B. Lippincott Company and Hemisphere Publishing Corporation, subsidiaries of Harper & Row, Publishers, Inc., New York.

1 2 3 4 5 6 7 8 9 0 BRBR 8 9 8

Library of Congress Cataloging-in-Publication Data

Purcell, Robert Y.
Handbook of control technologies for hazardous air pollutants / Robert Y. Purcell, Gunseli Sagun Shareef; edited and adapted by JACA Corporation for the Environmental Protection Agency.
p. cm.
Rev. ed. of: Evaluation of control technologies for hazardous air pollutants. 1986.
Includes bibliographies and index.
ISBN 0-8911-825-7
1. Air—Pollution—United States—Handbooks, manuals, etc. 2. Pollutants—Environmental aspects—United States—Handbooks, manuals, etc. 3. Hazardous substances—Environmental aspects—United States—Handbooks, manuals, etc. I. Shareef, Gunseli Sagun. II. Purcell, Robert Y. Evaluation of control technologies for hazardous air pollutants. III. JACA Corporation. IV. United States. Environmental Protection Agency. V. Title.
TD883.1.P87 1988 87-28081
604.7—dc19 CIP
ISBN 0-89116-825-7

Contents

Figures

Figures (continued)

Tables

Tables (continued)

Foreword

The purpose of this Handbook is to help technical personnel select, evaluate, and cost air pollution control techniques for reducing or eliminating the emission of potentially hazardous air pollutants from industrial and commercial sources. It should be useful for preparing and reviewing permit applications and for informing concerned parties of the type, basic design, and cost of available control systems.

Today's rapidly changing industrial technologies, products, and practices frequently carry with them an increasing generation of hazardous air pollutants (HAPs). When energy and material resources are extracted, processed, converted, and used, the consequent impacts on health and the environment require that efficient pollution control methods be used.

The 1970 amendments to the Clean Air Act require the U.S. Environmental Protection Agency (EPA) to set National Ambient Air Quality Standards for criteria air pollutants found throughout the country. The Clean Air Act Amendments also require EPA to review and regulate hazardous air pollutants, defined as those air pollutants that can contribute to increased mortality or serious illness but which are not already regulated as criteria pollutants.

Since the definition of a HAP is very broad and encompasses thousands of specific compounds, it is not practical to develop and all-inclusive list of HAP compounds and compound-specific control techniques. However, the number of generic air pollution control techniques available is small, and the factors affecting the cost and performance of these controls as applied to many noncriteria pollutants have been identified and discussed in the literature.

This Handbook has been reviewed in accordance with the U.S. Environmental Protection Agency's peer and administrative review policies and approved for publication. Mention of trade names or commercial products does not constitute endorsement or recommendation for use.

Acknowledgments

This handbook has been adapted from a two-volume EPA report titled *Evaluation of Control Technologies for Hazardous Air Pollutants*. The report was written by Robert Y. Purcell, Pacific Environmental Services Inc., Durham, North Carolina, and Gunseli Sagun Shareef, Radian Corporation, Research Triangle Park, North Carolina. They were assisted by: Vishnu S. Katari, Karin C. C. Gschwandtner, Michael K. Sink, and Charlotte R. Clark of Pacific Environmental Services Inc.; and Andrew J. Miles, D. Blake Bath, and Glynda E. Wilkins of the Radian Corporation. Dr. Bruce A. Tichenor of EPA's Air and Energy Engineering Research Laboratory, Research Triangle Park, North Carolina, was the EPA project officer for the report.

This handbook was adapted and produced by JACA Corporation, Fort Washington, Pennsylvania. Norman Kulujian of EPA's Center for Environmental Research Information, Cincinnati, Ohio, was the EPA project officer for this publication.

Nomenclature[a]

a = packing constant
A = heat exchanger surface area, ft^2
A_{bed} = carbon bed cross sectional area, ft^2
A_{column} = absorber column cross sectional area, ft^2
A_{con} = condenser surface area, ft^2

A_{nc} = net cloth area, ft^2
A_p = collection plate area, ft^2
A_t = venturi scrubber throat area, ft^2
A_{tc} = total cloth area, ft^2
ABS = abscissa (Figure 4.7-2)

AC = adsorption capacity of carbon bed, lb HAP/100 lb carbon
A/C = air to cloth ratio for baghouse, acfm/ft^2
AF = absorption factor
b = packing constant
c = packing constant

C = annual credits, $/yr
C_{req} = amount of carbon required, lb
Cp_{air} = average specific heat of air, Btu/scf-°F
$\overline{Cp}_{air}$ = average specific heat of air, Btu/lb-mole-°F
Cp_{com} = average specific heat of combined gas stream, Btu/scf-°F

$\underline{Cp}_{coolant}$ = average specific heat of coolant, Btu/lb-°F
Cp_e = average specific heat of emission stream, Btu/scf-°F
$\underline{Cp}_e$ = average specific heat of emission stream, Btu/ lb-°F
$\underline{Cp}_f$ = average specific heat of supplementary fuel (natural gas), Btu/lb-°F
Cp_{fg} = average specific heat of flue gas, Btu/scf-°F

$\underline{Cp}_{flg}$ = average specific heat of flare gas, Btu/lb-°F
$\underline{Cp}_w$ = average specific heat of water, Btu/lb-°F
$\overline{Cp}_{HAP}$ = average specific heat of HAP, Btu/lb-mole-°F
CE = collection efficiency (based on mass), %
CRF = capital recovery factor

CRF_w = weighted average capital recovery factor
d = packing constant
D = annual direct labor costs, $/yr
D_{bed} = carbon bed diameter, ft
D_{column} = absorber column diameter, ft

D_{duct} = duct diameter, in
D_p = mean particle diameter, μm
D_t = venturi scrubber throat diameter, ft
D_{tip} = flare tip diameter, in
D_G = diffusivity in gas stream, ft^2/hr

[a]English units are used throughout this report. Many engineering handbooks provide conversion factors for English to metric units.

D_L	=	diffusivity in liquid, ft^2/hr
D_1	=	annual operating labor cost, \$/yr
D_2	=	annual supervision labor cost, \$/yr
DE	=	destruction efficiency, %
$DE_{reported}$	=	reported destruction efficiency, %
DP	=	stream dew point, °F
ex	=	excess air, % (volume)
f	=	fraction
FE	=	fabricated equipment cost index
FER	=	fan electricity requirement, kWh
g	=	packing constant
g_c	=	gravitational constant, = 32.2 ft/sec^2
G	=	gas (emission stream) flow rate, lb/hr
G_{area}	=	gas (emission stream) flow rate based on column cross sectional area, $lb/sec\text{-}ft^2$
$G_{area,f}$	=	gas (emission stream) flow rate at flooding conditions based on column cross sectional area, $lb/sec\text{-}ft^2$
G_{mol}	=	gas (emission stream) flow rate, lb-mole/hr
h_d	=	heat content of emission stream after dilution, Btu/scf
h_e	=	heat content of emission stream, Btu/scf
h_f	=	lower heating value of supplementary fuel (natural gas), Btu/scf
h_{flg}	=	flare gas heat content, Btu/scf
ΔH	=	heat of vaporization of HAP, Btu/lb-mole
H_{con}	=	enthalpy change associated with condensed HAP, Btu/min
H_f	=	supplementary heat requirement (heat supplied by the supplementary fuel), Btu/min
H_{load}	=	condenser heat load, Btu/hr
H_{noncon}	=	enthalpy change associated with noncondensible vapors, Btu/min
H_{uncon}	=	enthalpy change associated with uncondensed HAP, Btu/min
H_G	=	height of a gas transfer unit, ft
H_L	=	height of a liquid transfer unit, ft
H_{OG}	=	height of a gas transfer unit (based on overall gas film coefficients), ft
Ht_{column}	=	absorber column packed height, ft
Ht_{total}	=	absorber column total height, ft
HAP_{con}	=	quantity of HAP condensed, lb-mole/min
HAP_e	=	inlet HAP concentration, ppmv
$HAP_{e,m}$	=	quantity of HAP in the emission stream entering the condenser, lb-mole/min
HAP_o	=	outlet HAP concentration, ppmv
$HAP_{o,m}$	=	quantity of HAP in the emission stream exiting the condenser, lb-mole/min
HP	=	fan power requirement, hp (horsepower)
HR	=	heat recovery in the heat exchanger, %
HRS	=	number of hours of operation per year
L	=	solvent flow rate, lb/hr
L"	=	solvent flow rate based on absorber column cross sectional area, $lb/hr\text{-}ft^2$
L_{gal}	=	solvent flow rate, gal/min
L_{mol}	=	solvent flow rate, lb-mole/hr
L_v	=	liquid flow rate in venturi scrubber, gal/min
$L_v/Q_{e,a}$	=	liquid to gas ratio, gal/10^3 acf
LEL	=	lower explosive limit, % (volume)
m	=	slope of the equilibrium curve
M	=	annual maintenance costs, \$/yr
M_e	=	moisture content of emission stream, %(volume)

M_1	=	annual maintenance labor cost, \$/yr
M_2	=	annual maintenance supervision cost, \$/yr
M_3	=	annual maintenance materials cost, \$/yr
MW_{avg}	=	average molecular weight of a mixture of components, lb/lb-mole
MW_e	=	average molecular weight of emission stream, lb/lb-mole
MW_{flg}	=	average molecular weight of flare gas, lb/lb-mole
$MW_{solvent}$	=	molecular weight of solvent, lb/lb-mole
MW_{HAP}	=	molecular weight of HAP (average molecular weight if a mixture of HAPs is present), lb/lb-mole
N	=	number of carbon beds
N_{OG}	=	number of gas transfer units (based on overall gas film coefficients)
O_2	=	oxygen content of emission stream, % (volume)
ORD	=	ordinate (Figure 4.7-2)
ΔP	=	total pressure drop for the control system, in H_2O
ΔP_a	=	absorber column pressure drop, lb/ft^2-ft
P_e	=	emission stream pressure, mm Hg
$P_{partial}$	=	partial pressure of HAP in emission stream, mm Hg
P_{vapor}	=	vapor pressure of HAP in emission stream, mm Hg
ΔP_{total}	=	absorber column total pressure drop, in H_2O
ΔP_v	=	pressure drop across venturi, in H_2O
PC	=	purchased equipment cost, \$
Q_a	=	flow rate of gas stream at actual conditions, acfm
Q_c	=	combustion air flow rate, scfm
Q_{com}	=	flow rate of combined gas stream entering the catalyst bed, scfm
$Q_{coolant}$	=	coolant flow rate, lb/hr
$Q_{cool,w}$	=	cooling water flow rate, lb/min
$Q_{e,a}$	=	emission stream flow rate at actual conditions, acfm
$Q_{e,s}$	=	saturated emission stream flow rate, acfm
Q_f	=	supplementary fuel (natural gas) flow rate, scfm
Q_{fg}	=	flue gas flow rate, scfm
$Q_{fg,a}$	=	flue gas flow rate at actual conditions, acfm
Q_{flg}	=	flare gas flow rate, scfm
$Q_{flg,a}$	=	flare gas flow rate at actual conditions, acfm
Q_{rec}	=	quantity of HAP recovered, lb/hr
Q_s	=	steam flow rate, lb/min
Q_w	=	cooling water flow rate, gal/min
r	=	packing constant
R	=	gas constant, = 0.73 ft^3-atm/lb-mole °R; = 1.987 cal/g-mole °K
R_{hum}	=	relative humidity, %
Ref	=	refrigeration capacity, tons
RE	=	removal efficiency, %
$RE_{reported}$	=	reported removal efficiency, %
s	=	packing constant
S	=	annual cost of operating supplies, \$/yr
Sc_G	=	Schmidt number for HAP/emission stream
Sc_L	=	Schmidt number for HAP/solvent system
St	=	steam ratio, lb steam/lb carbon
SV	=	space velocity, hr^{-1}
t_c	=	cleaning interval, min

t_r = residence time, sec
T = temperature, °F
T_c = combustion temperature, °F
T_{ci} = temperature of combined gas stream entering the catalyst bed, °F
T_{co} = temperature of flue gas leaving the catalyst bed, °F

T_{con} = condensation temperature, °F
$T_{cool,i}$ = inlet temperature of coolant, °F
$T_{cool,o}$ = outlet temperature of coolant, °F
T_e = emission stream temperature, °F
$T_{e,s}$ = temperature of saturated emission stream, °F

T_{flg} = flare gas temperature, °F
T_{he} = emission stream temperature after heat exchanger, °F
T_r = reference temperature, = 70 °F
T_{sti} = inlet steam temperature, °F
T_{sto} = condensed steam outlet temperature, °F

T_{wi} = inlet cooling water temperature, °F
T_{wo} = outlet cooling water temperature, °F
ΔT_{LM} = logarithmic mean temperature difference, °F
Th_{column} = absorber column thickness, ft
U = overall heat transfer coefficient, Btu/hr-ft^2- °F

U_d = drift velocity of particles, ft/sec
U_{duct} = velocity of gas stream in the duct, ft/min
U_e = emission stream velocity through carbon bed, ft/min
$U_{e,s}$ = throat velocity of saturated emission stream, ft/sec
U_{flg} = flare gas exit velocity, ft/sec
U_{max} = maximum flare gas velocity, ft/sec
Ut = annual utility costs, \$/yr
V_c = combustion chamber volume, ft^3
V_{carbon} = volume of carbon bed, ft^3
V_{bed} = catalyst bed requirement, ft^3
$V_{packing}$ = absorber column packing volume, ft^3
W = particle grain loading, gr/acf
Wt_{column} = absorber column weight, lb
$\bar{x}$ = mole fraction of solute in solvent, moles solute/(moles solute + moles solvent)
$\bar{X}$ = mole fraction of gaseous component in liquid, moles solute/ moles solvent

$\bar{y}$ = mole fraction of solute in air, moles solute/(moles solute + moles air)
Y = packing constant
$\bar{Y}$ = mole fraction of solute in air, moles solute/moles air

Z_{bed} = carbon bed depth, ft
ϵ = packing constant
λ = latent heat of vaporization for steam, Btu/lb
η = fan efficiency, percent
ρ_{bed} = density of carbon bed, lb/ft^3
ρ_c = density of carbon steel plate, lb/ft^3
ρ_G = density of gas (emission stream), lb/ft^3

ρ_L = density of solvent, lb/ft^3
θ_{ads} = cycle time for adsorption, hr
θ_{reg} = cycle time for regeneration, hr
$\theta_{dry\text{-}cool}$ = cycle time for drying and cooling the bed, hr
μ_L = viscosity of solvent, centipoise
μ_L'' = viscosity of solvent, lb/ft-hr

Chapter 1
Introduction

1.1 Objective

The objective of this handbook is to present a methodology for determining the performance and cost of air pollution control techniques for reducing or eliminating the emission of potentially hazardous air pollutants (HAP's) from industrial/commercial sources. (Note: The term "hazardous" in this document is very broad. It is not limited to the specific compounds listed under current regulations [i.e., the Clean Air Act, the Resource Conservation and Recovery Act, and the Toxic Substances Control Act].) This handbook is to be used by EPA regional, State, and local air pollution control agency technical personnel for two basic purposes: (1) to respond to inquiries from interested parties (e.g., prospective permit applicants) regarding the HAP control requirements that would be needed at a specified process or facility, and (2) to evaluate/review permit applications for sources with the potential to emit HAP's. It should be noted that this document provides general technical guidance on controls and does not provide guidance for compliance with specific regulatory requirements for hazardous air pollutants. Specifically, it does not specify design requirements necessary to achieve compliance with standards established under specific programs such as Section 112 of the Clean Air Act or standards established under the Resource Conservation and Recovery Act. Such requirements vary with the hazardous air pollutant emitted and with the emission source; thus, regulatory-specific detailed specifications are beyond the scope of this handbook.

Section 1.2 discusses the use of this handbook. Chapter 2 assists the user in identifying HAP's and their respective emission sources. Chapter 2 also identifies the key emission stream characteristics necessary to select appropriate control techniques. Chapter 3 provides additional information to assist the user in the control technique selection process for each HAP emission source/stream. Chapter 4 presents simple step-by-step procedures to determine basic design parameters of the specific control devices and auxiliary equipment. Chapter 5 provides the necessary data and procedures to determine order-of-magnitude estimates (-60 to $+30$ percent) for the capital and annualized costs of each control system. Appendices A and B present supplementary data and calculation procedures. Appendix C contains blank worksheets to be used while performing the functions described in this handbook. These worksheets are masters from which to make copies.

Additional appendices can be found in *Evaluation of Control Technologies for Hazardous Air Pollutants—Appendices* (EPA 600/7-86-009b; NTIS order no. PB 86-167/038/AS; $22.95, price subject to change). The additional appendices further clarify and expand the text and give derivations of equations, calculation procedures, and unit conversion techniques. They are referenced in this handbook. The volume can be ordered from the National Technical Information Service (NTIS), 5285 Port Royal Road, Springfield, Virginia 22161; (703) 487-4650.

A good source of current information pertaining to HAP's is the "Air Toxics Information Clearinghouse," which was established by EPA in response to State and local agency requests for assistance in the exchange of information on toxic air pollutants. The Clearinghouse is operated by EPA's Office of Air Quality Planning and Standards (OAQPS) in close coordination with the State and Territorial Air Pollution Program Administrators (STAPPA) and the Association of Local Air Pollution Control Officials (ALAPCO). The Clearinghouse collects and disseminates information from State and local agencies, as well as making users aware of air toxics information available from EPA and other Federal agencies. Specifically, the Clearinghouse collects the following air toxic information from State and local agencies: regulatory program descriptions, acceptable ambient concentrations on ambient standards, toxic pollutant research, source permitting, ambient monitoring, toxicity testing, and source testing.

The Clearinghouse provides an on-line data base containing all toxic-related information submitted by State and local agencies, bibliographic citations for relevant reports by EPA and other Federal agencies, and references for ongoing EPA air toxic projects. It also publishes a quarterly newsletter with articles on current air toxics concerns. Finally, the Clearinghouse periodically publishes various special reports on topics of interest to users. For further information regarding the "Air Toxics Information Clearinghouse," contact the appropriate EPA regional office air toxics contact, or EPA/OAQPS, Pollutant Assessment Branch, MD-12, Research Triangle Park, North Carolina 27711; (919) 541-5645 or FTS 629-5645.

1.2 How to Use the Handbook

Figure 1-1 is a flowchart of the steps performed when responding to inquiries; Figure 1-2 presents the same type of flowchart when reviewing permits. As shown by these figures, these two functions are basically the same; the only substantive difference is that the review process also compares the determined/calculated parameters with the corresponding parameters stated in the permit application to ensure that the control system(s) proposed by the applicant will provide the required reduction of HAP emissions.

Once an inquiry or permit application is received, determine the HAP's applicable to the source category in question (Section 2.2). The HAP's are categorized under four headings: organic vapor, organic particulate, inorganic vapor, and inorganic particulate. (Note: For each HAP group, a list of potentially or suspected hazardous compounds that may be emitted as a HAP from the source category is provided. This listing is neither all-inclusive nor a declaration that the compounds presented are hazardous.) Next, identify the potential emission sources for each HAP group (Section 2.2). The HAP emission sources are listed under one of three classifications: process point sources, process fugitive sources, and area fugitive sources. (Note: See Section 2.2 for classification definitions.) After each emission source is determined, identify the key HAP emission stream characteristics (e.g., HAP concentration, temperature, flow rate, heat content, particle size) needed to select the appropriate control technique(s) (Section 3.2). Obtain the actual values for these characteristics from the owner/operator or from available literature if the owner/operator cannot provide the necessary data. If two or more emission streams are combined prior to entry into an air pollution control system, determine the characteristics of the combined emission stream (Appendix B.1).

Depending upon the specific regulation and the type/characteristics of the HAP emission source/stream, the remaining steps in the methodology will differ. There are four basic "formats" for a regulation: (1) a particular "control device" may be required, (2) a "numerical limit" may be specified, (3) a "technology forcing" requirement may be imposed, and (4) a specific work practice or "other" related practice may be required. The regulation format will define the steps that lead to the selection of the appropriate control technique(s). The "control device" and "other" formats specify the appropriate control technique(s). A "numerical limit" format requires the determination of the HAP removal efficiency before the appropriate control technique(s) can be identified. Lastly, the "technology forcing" format has two paths: one where the cost of the control system is a factor in the decision, and one where cost is not a factor. If control system cost is a factor, the agency must determine the cost constraints that will be imposed on the control technique selection process. The steps that occur in defining the HAP control requirements will depend upon each agency's regulatory policies.

The HAP emission stream characteristics, in conjunction with the limitations imposed by the applicable regulations, are used to select the appropriate control techniques (Chapter 3) for each HAP emission source/stream. General guidelines are provided that match specific control devices with specified emission stream properties (e.g., HAP content, temperature, moisture, heat content, particle size, flow rate). Basic design parameters are then determined to provide general design conditions that should be met or exceeded for each selected control technique to achieve the specified HAP removal efficiency (Chapter 4). This exercise also identifies which of the selected control techniques will not achieve the desired HAP control requirements. The basic design parameters also can be used to obtain an order-of-magnitude cost estimate for each control device (Chapter 5). As noted above, this cost information can be an integral part of the HAP control system selection process. After completing the above process, a HAP control program can be recommended or evaluated. (For an example of a State HAP program, see Appendix A.1, reference 1.)

Example Case

To guide the user through the steps and calculations described in this handbook, examples are provided throughout the text. As shown here, each example is boxed. The primary example case pertains to a hypothetical plant owner requesting assistance in determining the type of control system that should be used on an emission stream generated by a paper coating drying oven. This example is carried through the entire handbook. Additional example emission streams are introduced in Chapter 3 to illustrate fully the control technique selection process and to clarify the design procedures of Chapter 4.

1.3 References

1. U.S. EPA. *Evaluation of Control Technologies for Hazardous Air Pollutants — Appendices.* EPA-600/7-86-009b (NTIS PB 86-167/038/AS). October 1985.

Figure 1-1. Steps used when responding to inquiries.

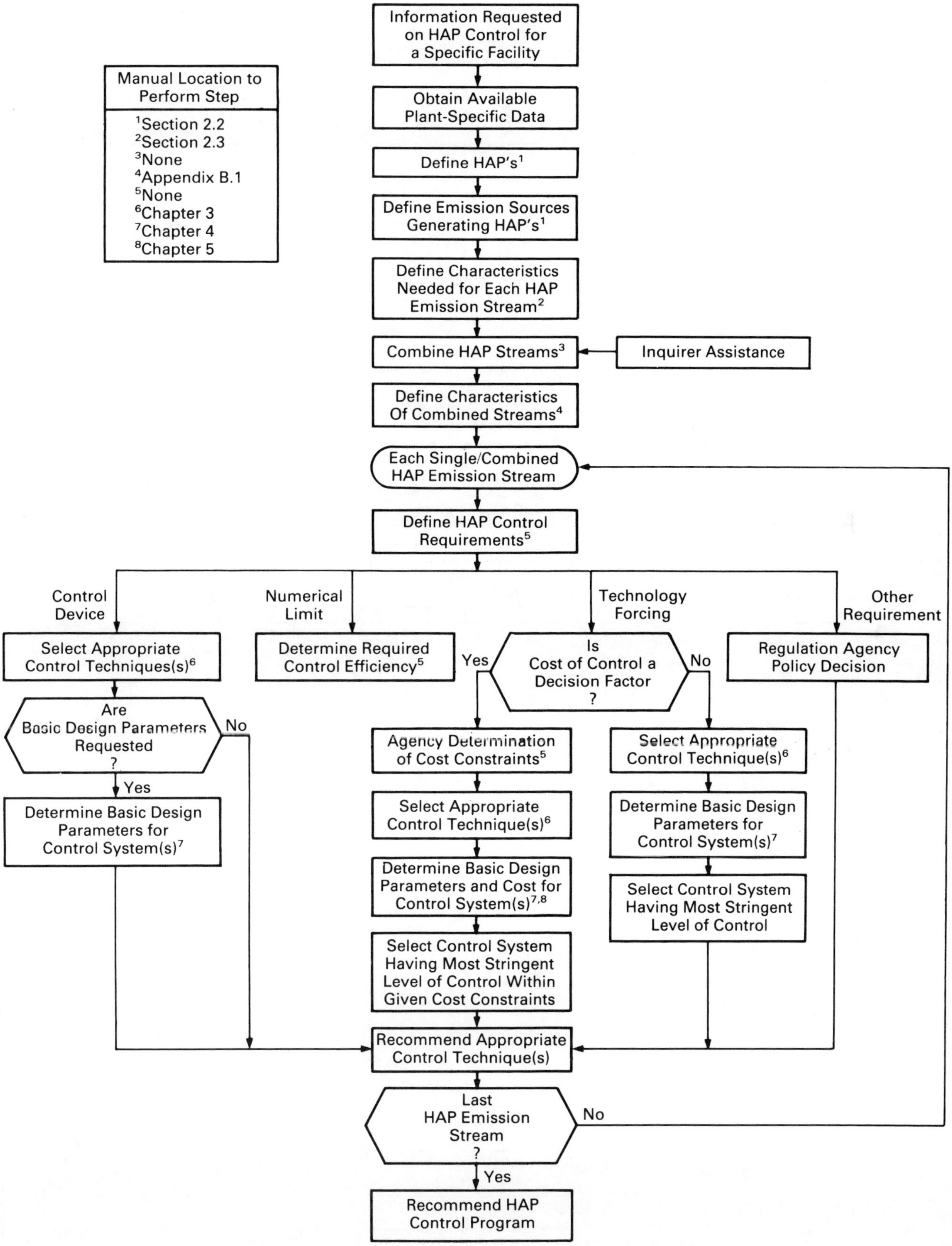

Figure 1-2. Steps used when reviewing permits.

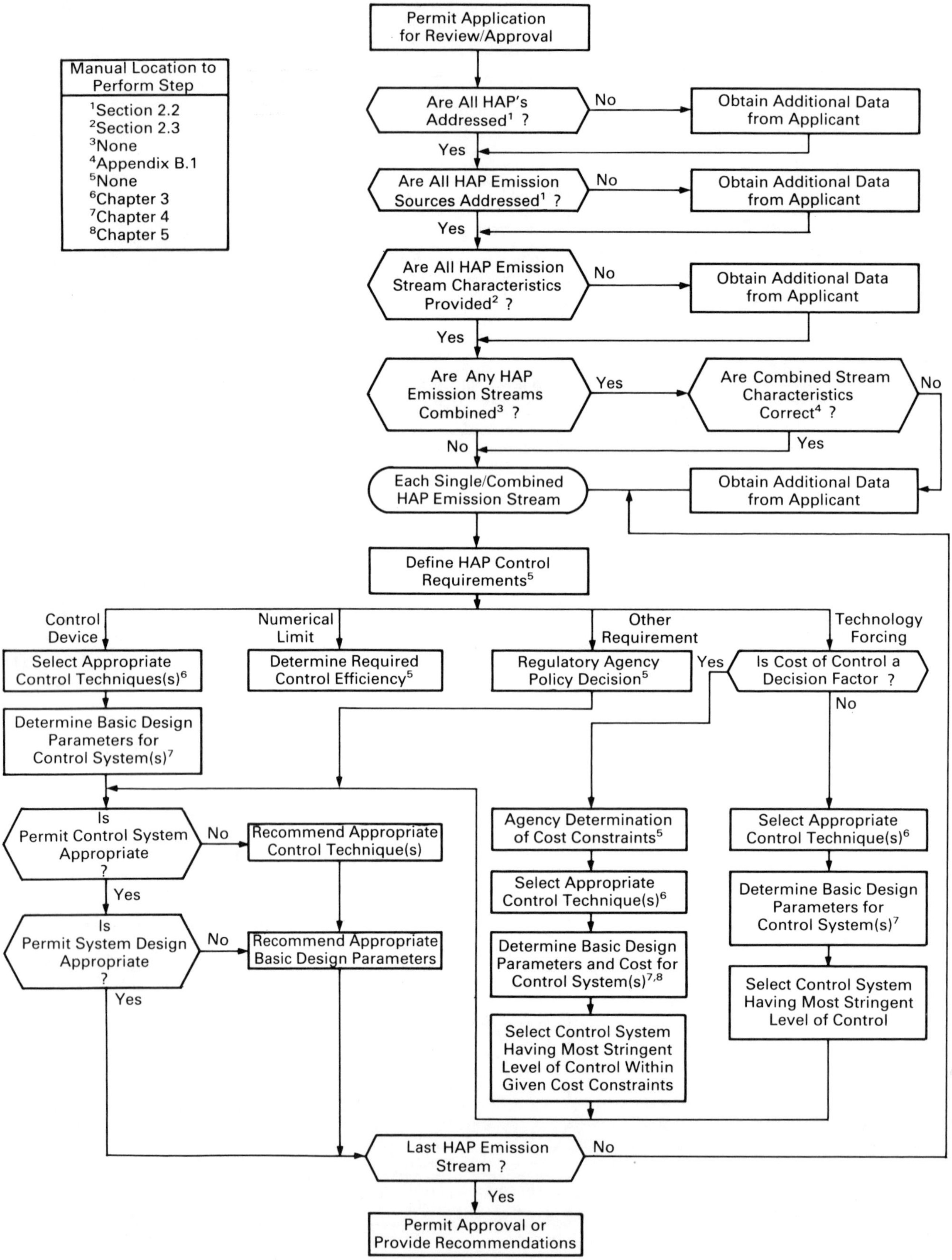

Chapter 2

HAP Emissions and Their Key Physical Properties

2.1 Background

This chapter's primary goal is to identify the following: (1) potential HAP's for a given source category and the specific sources that may emit the potential HAP's—Section 2.2, and (2) key emission stream physical properties needed to select appropriate control strategies and size control devices for the HAP emission sources Section 2.3. Specific source categories are divided into nine general classifications in this manual. (Note: The general classification system is a hybrid of the classification systems used by the following documents: (1) U.S. EPA, *Compilation of Air Pollutant Emission Sources—Third Edition: Supplements 1-15*, AP-42, January 1984; and (2) the U.S. EPA, *BACT/LAER Clearinghouse—A Compilation of Control Technology Determinations*, April 1983.) Every possible source category cannot be listed; however, similarities exist between many categories. Thus, the user should be able to obtain some guidance for any specific facility. Only those source categories that are known to emit potential HAP's are presented in this handbook. (For a listing of chemical hazard information profiles [CHIP's] and CAS numbers, see appendix A.2, reference 1.)

Individual source categories have been classified based on the manufacturing process associated with emissions of potential HAP's. The Solvent Usage Operations classification includes processes dependent on solvents, such as surface coating and dry cleaning operations. Metallurgical Industries include processes associated with the manufacture of metals, such as primary aluminum production. Processes and operations associated with the manufacture of organic and inorganic chemicals have been grouped into the Synthetic Organic and Inorganic Chemical Manufacturing classifications, respectively. Industries using chemicals in the formulation of products are classified as Chemical Products Industries. The Mineral and Wood Products Industries classifications include operations such as asphalt batch plants and kraft pulp mills, respectively. The Petroleum Related Industries classification is defined as oil and gas production, petroleum refining, and basic petrochemicals production. Combustion Sources are utility, industrial, and residential combustion sources using coal, oil, gas, wood, or waste-derived fuels.

To assist the user in recording the pertinent information, a worksheet has been provided. A copy of this worksheet, the HAP Emission Stream Data Form, is presented in Appendix C.1. An example of a partially completed worksheet is shown in Figure 2-1. This worksheet is designed to record information pertaining to one emission stream, be it a single stream or a combined stream consisting of several single streams.

Example Case

Information has been requested by a paper coating plant owner regarding the control of an emission stream from his facility's drying operations. The most likely generic classification to include a paper coating plant (one of many surface coating industries) would be Solvent Usage Operations. To determine if a paper coating facility is listed within this category, see Section 2.2.1 and Table 2-1, which indicates that the initial choice was correct (i.e., paper coating is listed under SC-Paper, Tapes, Labels) and the information retrieval process can begin.

2.2 Identification of Potential HAP's and Emission Sources

The purpose of this section is to present general information on emissions of potential HAP's by source category. Within each of the nine general classifications, information is presented on the types of potential HAP's that may be emitted by a particular source category. This information includes the names of specific compounds, the classification of the compounds (i.e., organic or inorganic), and the form in which these compounds would be emitted (i.e., vapor or particulate). Owing to process variations, actual emissions from specific facilities may differ from the general information presented. Complete identification of HAP emissions is best accomplished with assistance from the owner/operator of the facility. (See Appendix A.3, reference 1, for a listing of trade names and common synonyms for HAP's.)

This section also presents information pertaining to the sources (e.g., processes) within each specific source category that have the potential to emit

Figure 2-1. A partially completed HAP Emission Stream Data Form for one of six emission streams (#1) generated at a fictitious company.

HAP EMISSION STREAM DATA FORM*

Company	Glaze Chemical Company	Plant Contact	Mr. John Leake
Location (Street)	87 Octane Drive	Telephone No.	(999) 555-5024
(City)	Somewhere	Agency Contact	Mr. Efrem Johnson
(State, Zip)		No. of Emission Streams Under Review	6

A. Emission Stream Number/Plant Identification #1 / #3 Oven Exhaust

		(a)	(b)	(c)
B.	HAP Emission Source	paper coating oven	-	-
C.	Source Classification	process point	-	-
D.	Emission Stream HAP's	toluene	-	-
E.	HAP Class and Form	organic vapor	-	-
F.	HAP Content (1,2,3)**	73 ppmv	-	-
G.	HAP Vapor Pressure (1,2)	28.4 mm Hg @ 77°F	-	-
H.	HAP Solubility (1,2)	insoluble in water	-	-
I.	HAP Adsorptive Prop. (1,2)	provided	-	-
J.	HAP Molecular Weight (1,2)	92 lb/lb-mole	-	-

K.	Moisture Content (1,2,3)	2% vol.	P.	Organic Content (1)***	44 ppmv CH_4, 4 ppmv ot
L.	Temperature (1,2,3)	120°F	Q.	Heat/O_2 Content (1)	0.4 Btu/scf / 20.6% vol.
M.	Flow Rate (1,2,3)	15,000 scfm (max)	R.	Particulate Content (3)	
N.	Pressure (1,2)	atmospheric	S.	Particle Mean Diam. (3)	
O.	Halogen/Metals (1,2)	none / none	T.	Drift Velocity/SO_3 (3)	/

U. Applicable Regulation(s)

V. Required Control Level

W. Selected Control Methods

*The data presented are for an emission stream (single or combined streams) prior to entry into the selected control method(s). Use extra forms if additional space is necessary (e.g., more than three HAP's). and note this need.

**The numbers in parentheses denote what data should be supplied depending on the data in steps "C" and "E":
1 = organic vapor process emission
2 = inorganic vapor process emission
3 = particulate process emission

***Organic emission stream combustibles less HAP combustibles shown on Lines D and F.

HAP's. In this handbook, emission sources are broadly classified into three groups: process point sources, process fugitive sources, and area fugitive sources. Point sources are, in general, individually defined. Reactors, distillation columns, condensers, furnaces, and boilers are typical point sources which discharge emissions through a vent-pipe or stack. These sources can be controlled through the use of add-on control devices. Process fugitive sources, like process point sources, are individually defined. Emissions from these sources include dust, fumes, or gases that escape from or through access ports and feed or discharge openings to a process (e.g., the open top of a vapor degreaser). Process fugitive sources also can include vent fans from rooms or enclosures containing an emissions source (e.g., a vent fan on a dry cleaner). These sources can be controlled by add-on control devices once the emissions are captured by hooding, enclosures, or closed vent systems and then transferred to a control device.

Area fugitive sources are characterized by large surface areas from which emissions occur. In addition, process equipment such as pumps, valves, and compressors are considered area fugitive sources; emissions from these sources occur through leaks during process operation. Although these sources are small, they are usually found in large numbers dispersed over a wide area in a given process. Area fugitive sources also include large and undefined emission sources such as waste treatment lagoons, raw material storage piles, roads, etc.

The sources listed in this handbook generate emissions; however, a definitive statement as to whether they emit a HAP cannot be made. As in the case of identifying the potential HAP's emitted at a specific facility, communication with the owner/operator is useful in identifying each source that emits a HAP. The listings found in this section are not all-inclusive; a specific facility may have an emission producing operation that is not common to its industry and, thus, the source may not be included here.

Table 2-1. Potential HAP's and Emission Sources for Solvent Usage Operations

Source Category	Potential HAP's[a]				Potential Emission Sources		
	Organic		Inorganic				
	Vapor	Particulate	Vapor	Particulate	Process Point	Process Fugitive	Area Fugitive
Solvent Degreasing	x				C	A,B,D	K
Dry Cleaning	x				F	E,G,H,I	J,K
Graphic Arts[b]	x				F	L,M,N	
Waste Solvent Reclaiming	x				N	I	J,K
SC[c]-Flatwood Paneling[d]	x				O	L,P	
SC-Machinery[e]	x			x	O	Q,R	
SC-Appliances[f]	x			x	O	Q,R	
SC-Metal Furniture	x			x	O	Q,R	
SC-Auto/Truck[g]	x			x	O	S,R	
SC-Fabrics	x				O	D,K,Q,R,T	
SC-Cans[h]	x				O	Q,U	
SC-Paper, Tapes, Labels	x				O	B,I,Q,T	
Magnetic Tape Coating	x				F	I,Q,T	
SC-Electrical Insulation	x				O	Q	
SC-Marine Vessels[i]	x			x			Q
Vinyl & Acrylic Coatings[j]	x				O	L,P	
SC-Wood Furniture	x				O	S,R,T	
SC-Trans. Vehicles[k]	x			x		S,R	
Machine Lubricants	x						Q
Rubber Tire Manufacturing	x				F	I,V,W	

Source Key

A — bath evaporation	I — solvent storage	Q — application area
B — solvent transfer	J — pipes, flanges, pumps	R — flashoff area
C — ventilation	K — transfer areas	S — spray booth
D — waste solvent disposal	L — rollers	T — solvent/coating mixing
E — washer	M — ink fountains	U — quench area
F — drying	N — condenser	V — green tire spraying
G — still, filtration	O — oven	W — sidewall/tread end/undertread cementing
H — cooker	P — coaters	

[a]References 2, 3, 4, 5, 6, 7, 8, 9, 10, 11, 13, 14, 15, 18, 31, 32.
[b]Includes flexography, lithography, offset printing, and textile printing.
[c]SC: surface coating.
[d]Includes coating of other flat stock.
[e]Includes coating of misc. metal parts, machinery, and equipment.
[f]Includes all categories of appliances, large and small.
[g]Includes coating of automobiles and light-duty trucks.
[h]Includes surface coating of coils, cans, containers, and closures.
[i]Includes coating and maintenance of marine vessels.
[j]Includes vinyl, acrylic, and nitrocellulose coatings.
[k]Includes coating of trucks, buses, railroad cars, airplanes, etc.

2.2.1 Solvent Usage Operations

Solvent usage operations are defined as manufacturing processes that use solvents, including such processes as surface coating operations, dry cleaning, solvent degreasing, waste solvent reclaiming, and graphic arts. Table 2-1 lists source categories within this group of operations that have been identified as sources of volatile organic compound emissions that may include potential HAP's. As is shown by Table 2-1, all solvent usage operations generate organic vapor emissions. (Note: Some of the emission sources generate aerosols [i.e., organic particulate]; however, the aerosols evaporate in a short time and the emissions normally are controlled as a vapor. Therefore, Table 2-1 does not indicate the presence of organic particulates.) Due to the large number of potential HAP's associated with these types of operations, the format of Table 2-1 prohibits the inclusion of compound-specific data. Potential HAP's that may be emitted by sources in Table 2-1 are summarized in Appendix A.1; this appendix lists both specific compounds and classes of compounds that may be emitted by sources within the category. Appendix A.1 can be used to determine whether a particular solvent usage operation may emit a specific potential HAP or group of potential HAP's, as well as to determine all solvent use operations that may emit a particular potential HAP. Table 2-1 presents the emission sources that may emit potential HAP's.

2.2.2 Metallurgical Industries

The metallurgical industries can be broadly divided into primary, secondary, and miscellaneous metal production operations. The majority of this industry is covered under SIC Codes 331, 332, 333, 334

Example Case

As directed by Section 2.2.1, Appendix A.1 is used to determine the potential HAP's. The potential HAP's for paper coating operations are as follows:

Specific Compounds	*Generic Compounds*
toluene	mineral spirits
xylene	other aromatics
ethylene glycol	alcohols
acetone	cellusolves
methyl ethyl ketone	ketones
methyl isobutyl ketone	esters
ethyl acetate	

Upon reviewing data from the solvent vendor, the owner determined that only toluene is present in the solvent being evaporated by the ovens. Table 2-1 indicates that toluene is an organic compound, and it would be emitted as a vapor. This information is then listed on the HAP Emission Stream Data Form provided in Appendix C.1 (see Figure 2-1).

Also for the example case, the source (oven: a process point source) was identified. If the user were interested in the other emission-releasing processes at a paper coating operation, Table 2-1 indicates that the remaining sources include solvent transfer, solvent storage, application areas, and solvent/coating mixing.

and 336. The term primary metals refers to production of the metal from ore. The secondary metals industry includes the recovery of metal from scrap and salvage and the production of alloys from ingots. The miscellaneous subdivision includes industries with operations that produce or use metals for final products. Table 2-2 presents the potential HAP's for these industries and the industry-specific emission sources.

2.2.3 Synthetic Organic Chemical Manufacturing Industry (SOCMI)

The SOCMI is a large and diverse industry producing several thousand intermediate and end-product chemicals from a small number of basic chemicals. Most of the chemicals produced by this industry fall under SIC Code 286. Due to the complexity of the SOCMI, a general approach is used in this section to describe generic emission sources and specific emission source types. This approach is identical to the approach used by EPA in its efforts to develop new source performance standards for the SOCMI.

A large proportion of the emissions from the SOCMI occur as organic vapors. However, organic particulate emissions may be generated in some processes (usually during the manufacture of chemicals that exist as solids at ambient conditions). The emissions typically contain raw materials (including impurities) used in and intermediate and final products formed during the manufacturing process. Many of these emission streams may contain HAP's due to the great number of compounds manufactured in the SOCMI. (See Appendix A.5, reference 1, for further information on specific emissions for different SOCMI processes.)

Potential emissions from this industry can be described generically as follows:

(a) Storage and handling emissions
(b) Reactor process emissions
(c) Separation process emissions
(d) Fugitive emissions
(e) Secondary emissions (e.g., from waste treatment).

Emissions can potentially occur from raw materials and product storage tanks as working and breathing losses through vents. Emissions from handling result during transportation or transfer of the volatile organic liquids. Reactor processes and separation processes are the two broad types of processes used in manufacturing organic chemicals. Reactor processes involve chemical reactions that alter the molecular structure of chemical compounds. A reactor process involves a reactor, or a reactor in combination with one or more product recovery devices. Product recovery devices include condensers, adsorbers, and absorbers. Emissions from reactor processes occur predominantly from venting of inert gases from reactors and product recovery devices, or from the release of organic compounds that cannot be recovered economically. Typical emission sources in reactor processes include point sources (e.g., vents on reactors and product recovery devices), process fugitive sources such as disposal of bottoms from the reactor or the product recovery devices, and area fugitive sources (e.g., pumps, valves, sampling lines, and compressors).

Separation processes often follow reactor processes and divide chemical product mixtures into distinct fractions. Emissions from separation processes are associated primarily with absorption, scrubbing, and distillation operations. Other sepa-

Table 2-2. Potential HAP's and Emission Sources for Metallurgical Industries

Source Category	Potential HAP's[a]				Potential Emission Sources		
	Organic		Inorganic				
	Vapor	Particulate	Vapor	Particulate	Process Point	Process Fugitive	Area Fugitive
Primary Aluminum Production			*12*	*12*	A,I,J, M,N,R	H,K,D	N,Q,U,Z
Primary Cadmium Production				*9*	J,E	O,P	N,Z
Metallurgical Coke	*3,8,13,19, 21,23,26, 27,28,30*	*18*	*4,29*	*1,5,6,7,9 14,15,16 17,20,22*	B	C,O,X	N,D,Q,U
Primary Copper Smelting		*18*	*1,12*	*1,5,9,11 14,15, 20,24*	F,J,T	G,H,K,O,P,X	N,Q,U,W,Z
Ferroalloy Production		*18*		*9,10,11,14 16,17,22,24*	J	H,K,O,P	N,Q,W
Iron and Steel Production		*18*	*12*	*6,9,10,11 14,16,17,22,24*	B,J,V	C,H,K,O,X	D,N,Q,U,W,Z
Primary Lead Smelting		*18*	*1,12*	*1,5,9,11 14,15,20*	J,V	H,K,O,P	N,Q,U,W,Z
Primary Zinc Smelting		*18*	*1,12*	*1,9,11 14,15, 20,24*	E,J,T,S	O	N,Q,U,W,Z
Manganese Production		*18*		*16*	J	H,K,M,P	N,Q,Z
Nickel Production		*18*	*1,12*	*1,9,14 17,20,24*	A,I,J,M,T	P	N,Q,Z
Secondary Aluminum Operations			*12*	*12,17*	J	H,K,P	U
Secondary Copper Operations (Brass and Bronze Production)		*18*	*24*	*9,11,14 17,20,24*	J	H,K,P	U
Gray Iron Foundries	*2,3,13, 19,21,23*	*18*		*1,6,7,9,10,11 14,15,16,17 22,24,25*	J,Y	H,K,G,P	U
Secondary Lead Smelting		*18*		*1,14,16,20*	J	H,K,P	U ,Q
Steel Foundries				*1,10,7,16,17,25*	J,Y	G,H,K,P	U
Secondary Zinc Processing			*24*	*9,15,17,20,24*	J,E,S	H,K,L,P	U
Lead Acid Battery Production			*14*	*14*		O,P	
Cadmium-Nickel Battery Production				*9,14*	V	N,O	
Dry Battery Production				*16*		M,N,O	
Misc. Lead Products			*14*	*5,14*		G,O,P	

Pollutant Key

1— arsenic
2— acrolein
3— acetaldehyde
4— ammonia
5— antimony
6— barium
7— beryllium
8— benzene
9— cadmium
10— chromium
11— copper
12— fluoride
13— formaldehyde
14— lead
15— mercury
16— manganese
17— nickel
18— polycyclic organic matter (POM)
19— phenol
20— selenium
21— toluene
22— vanadium
23— xylene
24— zinc
25— iron
26— cresols
27— cyanides
28— pyridine
29— hydrogen sulfide
30— methyl mercaptan

Source Key

A — calciner
B — coke oven
C — coke oven charging/pushing
D — coke quenching
E — condenser
F — converter
G — converter charging/etc.
H — furnace tapping
J — furnace
K — furnace charging
L — galvanizing vessel
M — material crusher/mill
N — material storage and handling
O — material preparation
P — metal casting
Q — outdoor storage pile
R — reduction cell
S — retort
T — roaster
U — service road
V — sintering machine
W — slag dumping
X — vessel leakage
Y — foundry mold and core decomposition
Z — mining operations

[a] References 6, 16, 17, 19, 20, 21, 22, 23, 24, 25, 26, 28, 29, 30, 33, 34, 35.

ration processes that may contribute to emissions include drying, filtration, extraction, settling, crystalization, quenching, evaporation, ion exchange, dilution, and mixing/blending. One of the more commonly employed separation techniques is distillation. Depending on the type of distillation system used (i.e., vacuum or nonvacuum), typical emission points can include condensers, accumulators, hot wells, steam jet ejectors, vacuum pumps, and pressure relief valves. Emission points from a vacuum distillation system are shown in Figure 2-2.

Although fugitive emissions are listed as a separate group, they can occur from storage and handling, reactor processes, and separation processes. Area fugitive sources include groups of valves, pressure relief devices, pumps and compressors, cooling towers, open-ended lines, and sampling systems. Process fugitive sources include hotwells, accumulators, and process drains from reactors, product recovery devices, and separation equipment.

Table 2-3 presents information on specific emission points and emission source types for each of

Table 2-3. Emission Sources for the SOCMI[a]

	Potential Emission Sources (Specific)		
Generic Source Category	Process Point	Process Fugitive	Area Fugitive[b]
Storage and Handling		A	B,C,D
Reactor Processes	E,F	G	C,D,H,I,J,K
Separation Processes	F,L	G,M,N	K
Fugitives		G,M,N	B,C,D,H,I J,K,M,N,O

Source Key

A — storage, transfer, and handling
B — spills
C — valves
D — flanges
E — reactors
F — product recovery devices (absorber, adsorber, condenser)
G — process drains
H — pumps
I — compressors
J — sampling lines
K — pressure relief devices
L — separation devices (distillation column, absorber, crystalizer, dryer, etc.)
M — hotwell
N — accumulator
O — cooling tower

[a]References 12, 36, 37, 38, 39, 40, 41.
[b]Groups of small point sources (e.g., valves, compressors, pumps, etc.) at a SOCMI plant are considered as area fugitive sources in this handbook.

Figure 2-2. Potential emission points (shaded) for a vacuum distillation column using steam jet ejectors with barometric condenser.(40)

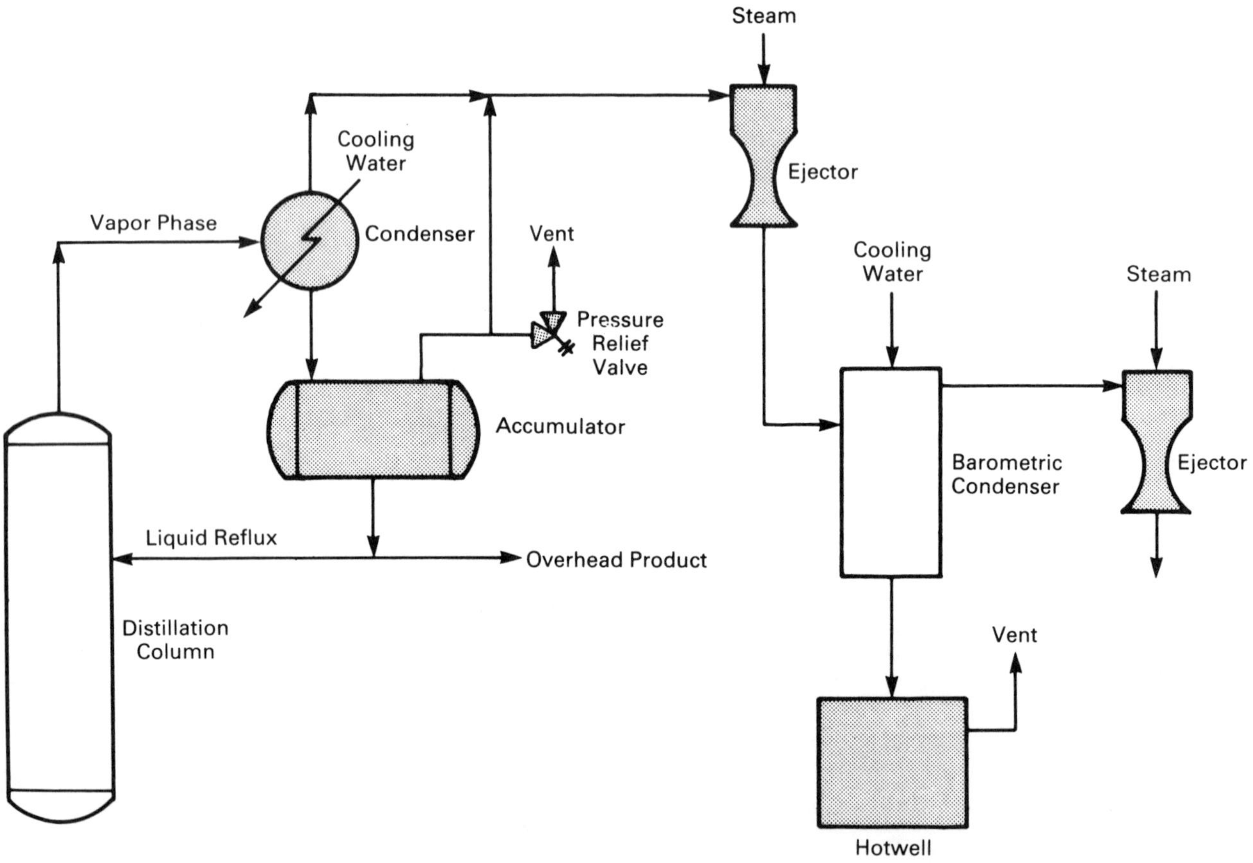

the generic emission source groups. Using this information, the user can identify the potential emission sources pertaining to his specific situation. (See Appendix A.5, reference 1, for an illustration of the approach outlined above and showing the emission sources and the HAP's potentially emitted from a SOCMI process.)

2.2.4 Inorganic Chemical Manufacturing Industry

This industry includes the manufacture of the basic inorganic chemicals before they are used in the manufacture of other chemical products. Most of the chemicals produced by this industry fall under SIC Code 281. Potential emissions from these processes may be high, but because of economic necessity they are usually recovered. In some cases, the manufacturing operation is run as a closed system, allowing little or no emissions to escape to the atmosphere. Table 2-4 presents the potential HAP's and industry-specific emission sources for these industries.

2.2.5 Chemical Products Industry

This industry includes the manufacture of chemical products such as carbon black, synthetic fibers, synthetic rubber and plastics, which may be used in further manufacture. Also included are the manufacture of finished chemical products for ultimate consumption such as pharmaceuticals, charcoal, soaps and detergents; or products to be used as materials or supplies in other industries such as paints, pesticides, fertilizers and explosives. Most of the chemical products are covered under SIC Codes 282, 283, 284, 285, 287 and 289. As in other chemical industries, the potential emissions from these processes may be high, but because of economic necessity they are usually recovered. Table 2-5 presents the potential HAP's and the industry-specific emission sources for this industry.

2.2.6 Mineral Products Industry

This industry involves the processing and production of various nonmetallic minerals. The industry includes cement production, coal cleaning and conversion, glass and glass fiber manufacture, lime manufacture, phosphate rock and taconite ore processing, as well as various other manufacturing processes. Most of this industry falls under SIC Codes 142, 144, 145, 147, 148, 149, 321, 322, 323, 324, 325, and 327. Table 2-6 presents the potential HAP's and the industry-specific emission sources industries.

2.2.7 Wood Products Industry

The wood products industry involves industrial processes that convert logs to pulp, pulpboard, hardboard, plywood, particleboard, or related wood products and wood preserving. This industry falls under SIC Codes 242, 243, 249, and 261. Chemical wood pulping involves the extraction of cellulose from wood by dissolving the lignin that binds the cellulose fibers together. The principal processes used in chemical pulping are kraft, sulfite, and neutral sulfite. Plywood production involves manufacturing wood panels composed of several thin wood veneers bonded together with an adhesive. The wood preserving process is one in which sawn wood products are treated by injection of chemicals that have fungistatic and insecticidal properties or impart fire resistance. Table 2-7 presents the potential HAP's and the industry specific emission sources for these industries.

2.2.8 Petroleum Related Industries

In this handbook, the petroleum related industries source category includes the oil and gas production industry, the petroleum refining industry, and the basic petrochemicals industry; these industries fall under SIC Codes 13 and 29.

The oil and gas production industry includes the following processes: exploration and site preparation, drilling, crude processing, natural gas processing, and secondary or tertiary recovery. The principal products of this industry are natural gas and crude oil.

The petroleum refining industry involves various processes that convert crude oil into more than 2,500 products, including liquefied petroleum gas, gasoline, kerosene, aviation fuel, diesel fuel, a variety of fuel oils, lubricating oils, and feedstocks for the petrochemicals industry. The different processes involved in the petroleum refining industry are crude separation, light hydrocarbon processing, middle and heavy distillate processing, and residual hydrocarbon processing.

In the basic petrochemicals industry, hydrocarbon streams from the oil and gas production and petroleum refining industries are converted into feedstocks for the organic chemical industry. These feedstocks include benzene, butylenes, cresol and cresylic acids, ethylene, naphthalene, paraffins, propylene, toluene, and xylene. The main processes used by this industry are separation, purification, and chemical conversion processes. (See Appendix A.6, reference 1, for a breakdown of typical processes involved in each of the three industries.)

Table 2-8 presents the potential HAP's that may be emitted from these industries. Table 2-9 provides more specific information on potential emissions from the petroleum refining industry segment of this generic category. A large proportion of the emissions occur as organic vapors; for example, benzene, toluene, and xylenes are the principal organic vapor emissions. This is due to the chemical

Table 2-4. Potential HAP's and Emission Sources for the Inorganic Chemical Manufacturing Industry

Source Category	Potential HAP's[a] Inorganic Vapor	Potential HAP's[a] Inorganic Particulate	Potential Emission Sources: Process Point	Process Fugitive	Area Fugitive
Aluminum Chloride	*4,10*		X	X	
Aluminum Fluoride	*17*		X	X	
Ammonia	*1*		B,D,E	K	J,S
Ammonium Acetate	*1*		X	X	
Ammonium — nitrate, sulfate thiocyanate, formate, tartrate	*1*		C,F,I,L	Q	
Ammonium Phosphate	*1,17*		X	X	
Antimony Oxide	*5*		X	X	
Arsenic — disulfide, iodide pentafluoride, thioarsenate tribromide, trichloride, trifluoride, trioxide orthoarsenic acid	*2*	*2*	H,U	K,Q,T	J,S
Barium — carbonate, chloride hydroxide, sulfate, sulfide		*6*	C,E,G,I,L,U	N,P,Q,T	
Beryllium — oxide, hydroxide		*7*	X	X	
Boric Acid and Borax		*9*	X	X	
Bromine	*8,10*		X	X	
Cadmium (pigment) — sulfide sulfoselenide, lithopone		*15*	X	X	
Calcium — carbide, arsenate phosphate	*3,17*	*2*	H	K,P	
Chlorine	*10*	*25*	H,C	K,R	J
Chlorosulfonic Acid	*19,24*		X	X	
Chromic Acid	*12*	*11,12*	H	K,N,O,Q	J,S
Chromium — acetate, borides halides etc.		*11*	X	X	
Chromium (pigment) — oxide		*11*	X	X	
Cobalt — acetate, carbonate halides, etc.		*13*	X	X	
Copper Sulfate	*14*		X	X	
Fluorine	*17*		X	X	
Hydrazine	*1,39*		X	X	
Hydrochloric Acid	*19,20*	*20*	B		
Hydrofluoric Acid	*17*		B,G	K,R	
Iodine (crude)	*10*	*38*	X	X	
Iron Chloride	*10,20*	*20*	X	X	
Iron (pigment) — oxide	*40*		X	X	
Lead — arsenate, halides hydroxides, dioxide, nitrate	*3*	*2,21*	G,L	P,Q	
Lead Chromate	*22*		G,R	P,Q	
Lead (pigments) — oxide carbonate, sulfate		*21*	G,R	P,Q	
Manganese Dioxide (potassium permanganate)	*24*	*23*	G,L	Q,P,T	
Manganese sulfate		*23*	G,L	Q,P,T	
Mercury — halides, nitrates, oxides		*25*	X	X	
Nickel — halides, nitrates, oxides		*26*		P,Q	
Nickel Sulfate	*27*	*26*	L	Q,T	
Nitric Acid	*28*	*28*	B,H	K,N,R	J,S
Phosphoric Acid	*10,17,2,30,18*	*30*			
Wet process			H,C,W	K,N,P,T	J,S
Thermal process			B,G	K,N,R,T	J,S
Phosphorus	*17*		X	X	
Phosphorus Oxychloride	*10*		X	X	

Source Category	Potential HAP's[a] Inorganic Vapor	Potential HAP's[a] Inorganic Particulate	Potential Emission Sources Process Point	Potential Emission Sources Process Fugitive	Potential Emission Sources Area Fugitive
Phosphorus Pentasulfide	*29,31*	*29*	X	X	
Phosphorus Trichloride	*32,10,29*	*29*	X	X	
Potassium — bichromate, chromate	*16*	*16*	I		
Potassium Hydroxide	*10*	*25*	X	X	
Sodium Arsenate		*2*	H	K,P	
Sodium Carbonate	*1*		I,L,V	P	
Sodium Chlorate	*10*		X	X	
Sodium Chromate — dichromate	*16*	*16*	G,I,L,M	P,Q	
Sodium Hydrosulfide	*18*		X	X	
Sodium — silicofluoride, fluoride	*17*		X	X	
Sulfuric Acid	*33,34*	*33*	A,B,C,H	K,R	J,S
Sulfur Monochloride — dichloride	*10*		X	X	
Zinc Chloride	*36,21*	*21*	X	X	
Zinc Chromate (pigment)	*35*		X	X	
Zinc Oxide (pigment)	*37*		X	X	

Pollutant Key
1— ammonia
2— arsenic
3— arsenic trioxide
4— aluminum chloride
5— antimony trioxide
6— barium salts
7— beryllium
8— bromine
9— boron salts
10— chlorine
11— chromium salts
12— chromic acid mist
13— cobalt metal fumes
14— copper sulfate
15— cadmium salts
16— chromates (chromium)
17— fluorine
18— hydrogen sulfide
19— hydrogen chloride
20— hydrochloric acid
21— lead
22— lead chromate
23— manganese salts
24— manganese dioxide
25— mercury
26— nickel
27— nickel sulfate
28— nitric acid mist
29— phosphorus
30— phosphoric acid mist
31— phosphorus pentasulfide
32— phosphorus trichloride
33— sulfuric acid mist
34— sulfur trioxide
35— zinc chromate
36— zinc chloride fume
37— zinc oxide fume
38— iodine
39— hydrazine
40— iron oxide

Source Key
A — converter
B — absorption tower
C — concentrator
D — desulfurizer
E — reformer
F — neutralizer
G — kiln
H — reactor
I — crystallizer
J — compressor and pump seals
K — storage tank vents
L — dryer
M — leaching tanks
N — filter
O — flakers
P — milling/grinding/crushing
Q — product handling and packaging
R — cooler (cooling tower, condenser)
S — pressure relief valves
T — raw material unloading
U — purification
V — calciner
W — hot well
X — no information

[a] References 6, 13, 20, 22, 23, 24, 25, 27, 28, 29, 30, 33, 42, 43, 44, 45, 46, 47, 48, 49, 50, 51.

Table 2-5. Potential HAP's and Emission Sources for the Chemical Products Industry

Source Category	Potential HAP's[a]				Potential Emission Sources		
	Organic		Inorganic		Process Point	Process Fugitive	Area Fugitive
	Vapor	Particulate	Vapor	Particulate			
Carbon Black	14,15,21 16,24	41	1	1,7,10,11 28,29,37	B,H	G,K,L	I
Charcoal	4,23,30	41			E		
Explosives	23				A,C,H	K	
Fertilizers	23,26,44		49		D,H,R,S,V	K,T	
Paint and Varnish	16,22, 31,46			6,28,43,48	N,O	L	
Pharmaceutical	17,18,31, 34,46			28	A,H,U,W	G,L	F
Plastics	23,33,35, 39,42,50, 51,52				A,P,V	K,L	F,I
Printing Ink	2,16,27, 42,46,45				Q		
Pesticides	9,16,17,18, 20,25,32, 36,39,47			28	A,H,O,X	G	F,I
Soap and Detergents	8,34	41	1	1,5	M,N,O	K,L	
Synthetic Fibers	3,8,13,14 19,23,32, 38,40,42, 46,24				A,H,J,O,U, V,X,Z	G,K	I
Synthetic Rubber	3,12,18,20, 22,33,34, 35,36, 46,49	41			A,H,O,P,X,Z	Y	F

Pollutant Key

1— arsenic
2— acrolein
3— acrylonitrile
4— acetic acid
5— boron
6— barium
7— beryllium
8— benzene
9— cresols
10— cadmium
11— chromium
12— chloroprene
13— caprolactum
14— carbon disulfide
15— carbonyl sulfide
16— carbon tetrachloride
17— chloroform
18— dichlorobenzene
19— dimethylformamide
20— dimethylamine
21— ethylene
22— ethylene dichloride
23— formaldehyde
24— hydrogen sulfide
25— hexachlorocyclopentadiene
26— hydrogen fluoride
27— ketones
28— mercury
29— manganese
30— methanol
31— methyl chloroform (1,1,1-trichloroethane)
32— maleic anhydride
33— 1,3- butadiene
34— morpholine
35— methylene chloride
36— nitrosomines
37— nickel
38— perchloroethylene
39— phosgene
40— phthalic anhydride
41— polycyclic organic matter
42— phenol
43— selenium
44— silicontetrafluoride
45— terpenes
46— toluene
47— xylene
48— zinc
49— ammonia
50— vinyl chloride
51— toluene diisocyanate
52— pyridine

Source Key

A — reactor
B — furnace
C — concentrator
D — neutralizer
E — kiln
F — compressor and pump seals; valves, flanges, open ended lines, sampling lines
G — storage tank vents
H — dryer
I — spills
J — spin cell or bath
K — product handling, finishing, and packaging
L — raw material transport and unloading
M — spray dryer
N — kettle
O — mixing tank (blend tank)
P — polymerization vessel
Q — cooking vessel
R — prill tower
S — granulator
T — screen
U — distillation
V — cooler (condenser)
W — crystallizer
X — filter
Y — milling/blending/compounding
Z — flash tank

[a] References 6, 13, 21, 22, 23, 33, 52, 53, 54, 55, 56, 57, 58, 59, 60.

Table 2-6. Potential HAP's and Emission Sources for the Mineral Products Industry

Source Category	Potential HAP's[a] Organic Vapor	Organic Particulate	Inorganic Vapor	Inorganic Particulate	Potential Emission Sources Process Point	Process Fugitive	Area Fugitive
Asbestos Products				*3,10,17*		D,N	I,L
Asphalt Batching Plants	*2,8,13*	*18*			B	F,J,M	I
Brick, Ceramic, and Related Clay Products			*10,12 21,26*	*7,10,12*	B,E,C	D,F,N	I,L
Refractories			*10,12*	*10,12*	B,E	D,F,N	I
Cement Manufacture		*18*	*10,12 17,23*	*7,9,10,14 15,17, 24,25*	E	F,G,N,S	I,L
Coal Cleaning (Dry Process)				*22*		M,N,R	I,L
Coal Cleaning (Wet Process)				*1,5,6,7,9, 10,11,14, 16,17,20, 21,24*	B,C	M,N	I,L
Coal Conversion	*8,19 27,28*	*18*	*4,23*	*1,5,7,9,14 16,17, 20,29*	B,H	F,G,M,N	I,L
Glass Fiber Manufacturing	*13,19*	*19*	*6,20,22*		C,O	D,F,G,N,P	I
Frit Manufacturing			*12*	*12*	B,C	S	I,L
Glass Manufacturing			*1,4,12, 14,26*	*1,5,6,12, 14,20,21*	C	D,F,M,N	I
Lime Manufacturing			*15*	*15*	E,T	G,R,S	I,L
Mercury Ore Processing			*15*	*15*	C	G,N	I,L
Mineral Wool Manufacturing	*13,19*		*12,23*		C,O	D,G,P	I,L
Perlite Manufacturing			*12*	*12*	B,C	G,M,N,S	I,L
Phosphate Rock Processing			*6,20,21*	*6,20,21*	A,B,Q	F,M,N,R	I,L
Taconite Ore Processing				*3*	C,Q	F,M,N,R	I,L

Pollutant Key
- *1*— arsenic
- *2*— aldehydes
- *3*— asbestos
- *4*— ammonia
- *5*— antimony
- *6*— barium
- *7*— beryllium
- *8*— benzene
- *9*— cadmium
- *10*— chromium
- *11*— copper
- *12*— fluoride
- *13*— formaldehyde
- *14*— lead
- *15*— mercury
- *16*— manganese
- *17*— nickel
- *18*— polycyclic organic matter (POM)
- *19*— phenol
- *20*— selenium
- *21*— boron
- *22*— coal dust
- *23*— hydrogen sulfide
- *24*— zinc
- *25*— iron
- *26*— chlorine
- *27*— cresols
- *28*— toluene
- *29*— phosphorus

Source Key
- A — calciner
- B — dryer
- C — furnace
- D — end-product forming and finishing
- E — kiln
- F — raw material preparation/mixing
- G — cooling
- H — reactor
- I — storage pile
- J — saturator
- L — mining operations
- M — raw material handling/transport
- N — raw material crusher/mill
- O — oven
- P — resin application
- Q — washers
- R — screening
- S — end-product handling/grinding/bagging
- T — hydrator

[a] References 6, 13, 20, 21, 22, 23, 24, 25, 26, 27, 28, 29, 45, 47, 64, 65, 66, 67, 68, 69, 70, 71, 73.

Table 2-7. Potential HAP's and Emission Sources for the Wood Products Industry

Source Category	Potential HAP's[a] Organic Vapor	Organic Particulate	Inorganic Vapor	Inorganic Particulate	Potential Emission Sources Process Point	Process Fugitive	Area Fugitive
Chemical Wood Pulping							
Kraft pulp mill	i	e	k	a,b,c,d	A,B,C,D		
Sulfite pulp mill	h	e	f,k	a,b,c,d	A,B,C,		
Neutral sulfite pulp mill		e	k	a,c,d	A,C,E		
Plywood, Particleboard, Hardboard	h,l,o,p				G	F	
Wood Preservative	j,g,m,n					F	

Pollutant Key
a — arsenic
b — asbestos
c — chromium
d — mercury
e — polycyclic organic matter (POM)
f — chlorine
g — chlorobenzene
h — formaldehyde
i — methyl mercaptan
j — dioxin
k — hydrogen sulfide
l — phenol
m — pentachlorophenol
n — cresols
o — abietic acid
p — pinene

Source Key
A — recovery furnace
B — digester
C — blow tank
D — lime kiln
E — fluidized bed reactor
F — resin and/or adhesive application
G — dryer

[a]References 4, 13, 21, 22, 23, 33, 56, 73.

Table 2-8. Potential HAP's for Petroleum Related Industries[a] (General Listing for Entire Source Category)

Potential HAP's Organic Vapor	Organic Particulate	Inorganic Vapor	Inorganic Particulate
Parafins (C_1-C_{10})	Coke fines	Sulfides (e.g., hydrogen sulfide, carbon disulfide, carbonyl sulfide)	Catalyst fines
Cycloparafins (C_6-C_9)			
Aromatics (e.g., benzene, toluene xylene)		Ammonia	
Phenols			
Sulfur containing compounds (e.g., mercaptans, thiophenes)			

[a]References 26, 75, 76, 77, 78, 79, 80.

composition of the two starting materials used in these industries—crude oil and natural gas. Crude oil is composed chiefly of hydrocarbons (paraffins, naphthalenes, and aromatics) with small amounts of trace elements and organic compounds containing sulfur, nitrogen, and oxygen. Natural gas is largely saturated hydrocarbons (mainly methane). The remainder may include nitrogen, carbon dioxide, hydrogen sulfide, and helium. Organic and inorganic particulate emissions, such as coke fines or catalyst fines, may be generated in some processes.

The emission sources within each of the petroleum related industries are given in Table 2-10. Sources of potential HAP emissions from the oil and gas production industry include blowouts during drilling operations; storage tank breathing and filling losses; wastewater treatment processes; and fugitive leaks in valves, pumps, pipes, and vessels. In the petroleum refining industry, potential HAP emission sources include distillation/fractionating columns, catalytic cracking units, sulfur recovery processes, storage tanks, fugitives, and combustion units (e.g., process heaters). Fugitive emissions are a major source of emissions in this industry. Emission sources in the basic petrochemicals industry are similar to those from the petroleum refining industry and the SOCMI (see Section 2.2.3).

2.2.9 Combustion Sources

The fuel combustion industry encompasses a large number of combustion units generally used to produce electricity, hot water, and process steam for industrial plants; or to provide space heating for industrial, commercial, or residential buildings. The combustion units may differ in size, configuration, and type of fuel burned. Coal, fuel oil, and natural gas are the major fossil fuels burned, although other fuels such as wood and various waste (e.g., waste oil) or byproduct fuels are burned in relatively small quantities. Industrial applications of both gasoline- and diesel-powered stationary internal combustion units such as generators, pumps, and well-drilling equipment are also included in this category.

The waste incineration category includes combustion processes whereby municipal solid wastes or sewage treatment sludges are disposed. Table 2-11 presents the potential HAP's and the facility specific emission sources for the above combustion sources.

Table 2-9. Potential HAP's for Petroleum Refining Industries (26) (Specific Listing for Petroleum Refining Segment)

Process	Potential HAP's: Organic Vapor	Organic Particulate	Inorganic Vapor	Inorganic Particulate
Crude Separation	a,b,d,e,f,g,h,i,j,k,l,m,o, A,B,C,D,E,F,J	o	c,m,t,u, v,x,y,L	p,I,Q,R
Light Hydrocarbon Processing	g,h,i,n,N,O,P	R	t,v	G,H,Q
Middle and Heavy Distillate Processing	a,d,e,f,g,h,i,j,k,l, F,J,K,O,P,S,T	o,R	m,t,u,v, x,y,L	p,q,G,H, I,Q,U
Residual Hydrocarbon Processing	a,d,e,f,g,h,i,j,k,l,n, F,J,M,N,P,S,T	o,R	m,s,t,u, v,x,y,L	p,q,G,H, I,Q,U
Auxiliary Processes	a,b,d,e,f,g,h,i,j,k,l,n, A,B,C,D,J,K,M,T	o,R	c,m,s,u, y,L	p,q,r,z, I

Pollutant Key

a — maleic anhydride
b — benzoic acid
c — chlorides
d — ketones
e — aldehydes
f — heterocyclic compounds (e.g., pyridines)
g — benzene
h — toluene
i — xylene
j — phenols
k — organic compounds containing sulfur (sulfonates, sulfones)
l — cresol
m — inorganic sulfides
n — mercaptans
o — polynuclear compounds (benzo pyrene, anthracene, etc.)
p — vanadium
q — nickel
r — lead
s — sulfuric acid
t — hydrogen sulfide
u — ammonia
v — carbon disulfide
x — carbonyl sulfide
y — cyanides
z — chromates
A — acetic acid
B — formic acid
C — methylethylamine
D — diethylamine
E — thiosulfide
F — methyl mercaptan
G — cobalt
H — molybdenum
I — zinc
J — cresylic acid
K — xylenols
L — thiophenes
M — thiophenol
N — nickel carbonyl
O — tetraethyl lead
P — cobalt carbonyl
Q — catalyst fines
R — coke fines
S — formaldehyde
T — aromatic amines
U — copper

Table 2-10. Emission Sources for Petroleum Related Industries

Source Category	Potential HAP Emission Sources: Process Point	Process Fugitive	Area Fugitive
Oil and Gas Production			
Exploration, Site Preparation and Drilling	A	C	D,E
Crude Processing	G	F,H	
Natural Gas Processing	G,J,K	H	I
Secondary and Tertiary Recovery Techniques	G		I
Petroleum Refining Industry			
Crude Separation	G,J,L	F,H,M,N	I
Light Hydrocarbon Processing	O,G	F,H	Q
Middle and Heavy Distillate Processing	G,O,P,R	F,H	I
Residual Hydrocarbon Processing	B,G,K,O,R	H	I
Auxiliary Processes	G	F,H	I
Basic Petrochemicals Industry			
Olefins Production	G,K,O	F,H	I
Butadiene Production	G,J,L,O,R	F,H,N	I
Benzene/Toluene/Xylene (BTX) Production	G,K,O,R	F,Q	I
Naphthalene Production	G,L,O	F,H	I
Cresol/Cresylic Acids Production	G,L	F,H	
Normal Paraffin Production	G,O	F,H	I

Source Key

A — blowout during drilling
B — visbreaker furnace
C — cuttings
D — drilling fluid
E — pipe leaks (due to corrosion)
F — wastewater disposal (process drain, blowdown, cooling water)
G — flare, incinerator, process heater, boiler
H — storage, transfer, and handling
I — pumps, valves, compressors, fittings, etc.
J — absorber
K — process vent
L — distillation/fractionation
M — hotwells
N — steam ejectors
O — catalyst regeneration
P — evaporation
Q — catalytic cracker
R — stripper

Table 2-11. Potential HAP's and Emission Sources for Combustion Sources

Source Category	Potential HAP's				Potential Emission Sources		
	Organic		Inorganic				
	Vapor	Particulate	Vapor	Particulate	Process Point	Process Fugitive	Area Fugitive
Coal Combustion	*14,25,28*	*19*	*1,2,8,9,13 17,27*	*1,2,5,6,8,9 10,11,15, 16,18,20, 22,24*	A,B	I	H
Oil Combustion	*14*	*19*	*13,17,27*	*1,2,5,6,8,9, 10,11,15, 16,18,22, 24,29*	A,B,E		
Natural Gas Combustion	*14*	*19*			A,B,E,F,		
Gasoline Combustion	*12,14*	*12,19*	*17*	*15*	G		
Diesel Combustion	*12*	*12,19*		*6,18*	G		
Wood Combustion	*3,4,12, 14,25*	*12,19*	*27*	*16,20*	A,B,C		
Waste Oil Combustion	*7,12,21, 23,26*	*12,19*		*6,8,9, 15,18*	A,B,D		
Municipal Refuse Incineration	*12*	*12,19*	*17,27*	*6,8,9,11, 15,16,18*	D		
Sewage Sludge Incineration	*12*	*12,19*	*17*	*1,6,8,9, 15,16,18*	D		
PCB Incineration	*12,21*	*12,19*		*30*	D,B		

Pollutant Key

1— arsenic
2— antimony
3— acetaldehyde
4— acetic acid
5— barium
6— beryllium
7— benzene
8— cadmium
9— chromium
10— cobalt
11— copper
12— dioxin
13— fluoride
14— formaldehyde
15— lead
16— manganese
17— mercury
18— nickel
19— polycyclic organic matter (POM)
20— phosphorus
21— polychlorinated biphenyls (PCB)
22— radionuclides
23— trichloroethylene
24— zinc
25— phenol
26— ethyl benzene
27— chlorine
28— pyridine
29— vanadium
30— dibenzofuran

Source Key

A — furnace
B — boiler
C — woodstove/fireplace
D — incinerator
E — gas turbine
F — reciprocating engine
G — industrial engine and/or equipment
H — coal storage pile
I — ash handling system

[a] References 6, 13, 20, 21, 23, 24, 25, 26, 27, 28, 29, 30, 33, 34, 35, 53, 72, 81, 82, 83, 84 .

2.3 Identification of Key Emission Stream Properties

This section identifies the emission stream physical properties needed to select the appropriate control technique(s) and to size the control device(s) for each identified HAP emission stream generated by a process source, be it either a process point source or a process fugitive source. Design and costing techniques for area fugitive emission control methodologies are outside the scope of this handbook; however, control techniques for vapor emissions and particulate emissions from area fugitive sources are discussed in Sections 3.2.4 and 3.3.2, respectively.

The actual/estimated values for the process emission stream properties should be obtained from the owner/operator or from available literature if the owner/operator cannot supply the necessary data. The values obtained are used in conjunction with the guidelines given in Chapter 3 to perform the control technique selection process. Table 2-12 lists the required information for organic vapor emissions, Table 2-13 for inorganic vapor emissions, and Table 2-14 for particulate emissions. After obtaining the values for the key physical properties for each HAP emission stream, record the data on the HAP Emission Stream Data Form found in Appendix C.1 (see Figure 2-1).

There will be occasions when it would be prudent for the owner/operator to combine similar emission streams. For example, if two or more emission streams require the use of the same control technique, it will normally be more cost effective to combine the streams and use just one control device as opposed to using a control device for each separate emission stream. If the owner/operator decides to combine emission streams, Appendix B.1 provides calculation procedures to determine the key effluent properties of combined emission streams.

Example Case

The emission stream from the coating oven has been identified as containing organic vapors. Therefore, the emission properties required are listed in Table 2-12. The necessary information was provided by the owner/operator; a source test was performed for a very similar operation at another plant owned by his company. The information is recorded on the worksheet presented in Appendix C.1. The source test provided the following data (see Figure 2-1):

HAP content	73 ppm (vol) toluene
Organic content	44 ppm (vol) methane 4 ppm (vol) other
Moisture content	2% (vol)
Halogen content	None
Metal content	None
Temperature	120°F
Pressure	Atmospheric
Flow rate	15,000 scfm (max)
Heat content	0.4 Btu/scf
Oxygen content	20.6% (vol)
HAP molecular weight	92 lb/lb-mole
HAP vapor pressure	28.4 mm Hg @ 77°F
HAP solubility	Insoluble in water

Table 2-12. Key Properties for Organic Vapor Emissions

Emission Stream Properties (Preferred units of measure)	HAP Properties[a]
HAP Content (ppm by volume)	Molecular Weight
Organic Content[b] (ppm by volume)	Vapor Pressure
Heat Content[c] (Btu/scf)	Solubility (graph)
Oxygen Content (% by volume)	Adsorptive Properties (isotherm plot)
Moisture Content (% by volume)	
Halogen/Metal Content (yes or no)	
Flow Rate (scfm)	
Temperature (°F)	
Pressure (mm Hg)	

[a]These properties pertain to the specific HAP or mixture of HAP's in the emission stream.
[b]Primary properties that affect control technique selection. Organic content is defined as organic emission stream combustibles less HAP emission stream combustibles.
[c]Heat content is determined from HAP/Organic Content (see Appendix B.1 for calculation procedures).

Table 2-13. Key Properties for Inorganic Vapor Emissions

Emission Stream Properties (Preferred units of measure)	HAP Properties[a]
HAP Content[b] (ppm by volume)	Molecular Weight
Moisture Content (% by volume)	Vapor Pressure
Halogen/Metal Content (yes or no)	Solubility (graph)
Flow Rate (scfm)	Adsorptive Properties (isotherm plot)
Temperature (°F)	
Pressure (mm Hg)	

[a]These properties pertain to the specific HAP or mixture of HAP's in the emission stream.
[b]Primary properties that affect control technique selection.

Table 2-14. Key Properties for Particulate Emissions

Emission Stream Properties (Preferred units of measure)	HAP Properties[a]
HAP Content (% by mass)	None
Particulate Content[b] (lb/acf)	
Moisture Content (% by volume)	
SO_3 Content (ppm by volume)	
Flow Rate (acfm)	
Temperature (°F)	
Particle Mean Diameter[c] (μm)	
Drift Velocity[c] (ft/sec)	

[a]These properties pertain to the specific HAP or mixture of HAP's in the emission stream.
[b]Data include total particulate loading and principle particulate constituent.
[c]These properties are necessary only for specific control techniques.

2.4 References

1. U.S. EPA. *Evaluation of Control Technologies for Hazardous Air Pollutants.* EPA-600/7-86-009 (NTIS PB 86-167/038/AS). October 1985.
2. National Paint and Coatings Association. *Section III: Paint and Coatings Markets.* Table A-6, Estimated Consumption of Solvents in Paints and Coatings, by Market - 1981. pp. 208-209. (no date).
3. U.S. EPA. *Organic Solvent Cleaners - Background Information for Proposed Standard* (Draft). EPA-450/2-78-045a. October 1979.
4. U.S. EPA. *End Use of Solvents Containing Volatile Organic Compounds.* EPA-450/3-79-032. May 1979.
5. U.S. EPA. *Source Assessment: Solvent Evaporation - Degreasing Operations.* EPA-600/2-79-019f. August 1979.

6. U.S. EPA. *Compilation of Air Pollutant Emission Sources. Third Edition: Supplements 1-15.* AP-42. January 1984.

7. U.S. EPA. *Guidance for Lowest Achievable Emission Rates for 18 Major Stationary Sources of Particulates, Nitrogen Oxides, Sulfur Dioxide, or Volatile Organic Compounds.* EPA-450/3-79-024. April 1979.

8. U.S. EPA. *Control of Volatile Organic Emissions from Existing Stationary Sources - Volume VI: Surface Coating of Miscellaneous Metal Parts and Products.* EPA-450/2-78-015. June 1978.

9. U.S. EPA. *Control of Volatile Organic Emissions from Existing Stationary Sources - Volume II: Surface Coating of Cans, Coils, Paper, Fabrics, Automobiles, and Light Duty Trucks.* EPA-450/2-77-008. May 1977.

10. U.S. EPA. *Control of Volatile Organic Compound Emissions from Large Petroleum Dry Cleaners.* EPA-450/3-82-009. September 1982.

11. U.S. EPA. *Pressure Sensitive Tape and Label Surface Coating Industry - Background Information for Proposed Standards.* EPA-450/3-80-003a. September 1980.

12. U.S. EPA. *Hazardous/Toxic Air Pollutant Control Technology: A Literature Review.* EPA-600/2-84-194. December 1984.

13. U.S. EPA. *Nonindustrial Sources of Toxic Substance Emissions and Their Applicability to Source Receptor Modeling.* Draft Report, EPA Contract No. 68-02-3509, Task No. 42. July 27, 1983.

14. U.S. EPA. *Control Technique Guidelines for the Control of Volatile Organic Emissions from Wood Furniture Coating* (Draft). April 1979.

15. U.S. EPA. *Flexible Vinyl Coating and Printing Operations - Background Information for Proposed Standards.* EPA-450/3-81-016a. January 1983.

16. U.S. EPA. *Background Information for New Source Performance Standards: Primary Copper, Zinc, and Lead Smelters - Volume 1: Proposed Standards.* EPA-450/2-74-002a. October 1974.

17. U.S. EPA. *Background Information for Standards of Performance: Electric Submerged Arc Furnaces for Production of Ferroalloys - Volume 1: Proposed Standards.* EPA-450/2-74-018a. October 1974.

18. U.S. EPA. *Control Techniques for Volatile Organic Compound Emissions from Stationary Sources - Third Edition* (Draft). April 1985.

19. U.S. EPA. *A Method for Characterization and Quantification of Fugitive Lead Emissions from Secondary Lead Smelters, Ferroalloy Plants, and Gray Iron Foundries*. EPA-450/3-78-003. January 1978.

20. U.S. EPA. *Locating and Estimating Air Emissions from Sources of Chromium*. EPA-450/4-84-007g. July 1984.

21. U.S. EPA. *A Survey of Emissions and Controls for Hazardous and Other Pollutants*. EPA-R4-73-021. February 1973.

22. U.S. EPA. *Industrial Sources of Hazardous Air Pollutants* (Draft). September 1983.

23. U.S. EPA. *Source Assessment: Noncriteria Pollutant Emissions (1978 Update)*. EPA-600/2-78-004T. July 1978.

24. U.S. EPA. *Locating and Estimating Air Emissions from Sources of Manganese* (Draft). September 1984.

25. U.S. EPA. *Locating and Estimating Air Emissions from Sources of Nickel*. EPA-450/4-84-007F. March 1984.

26. U.S. EPA. *Potentially Hazardous Emissions from the Extraction and Processing of Coal and Oil*. EPA-650/2-75-038. April 1975.

27. U.S. EPA. *Review of National Emission Standards for Mercury*. EPA-450/3-84-01. December 1984.

28. U.S. EPA. *Status Assessment of Toxic Chemicals: Lead*. EPA-600/2-79-210h. December 1979.

29. U.S. EPA. *Status Assessment of Toxic Chemicals: Mercury*. EPA-600/2-79-210i. December 1979.

30. U.S. EPA. *Sources of Copper Air Emissions*. EPA-600/2-85-046. April 1985.

31. U.S. EPA. *The Use and Fate of Lubricants, Oils, Greases and Hydraulic Fluids in The Iron and Steel Industry*. EPA-600/2-78-101. May 1978.

32. U.S. EPA. *Rubber Tire Manufacturing Industry - Background Information for Proposed Standards*. EPA-450/3-81-008a. July 1981.

33. U.S. EPA. *Human Exposure to Atmospheric Concentrations of Selected Chemicals*. EPA Contract No. 68-02-3066. February 1982.

34. U.S. EPA. *Survey of Cadmium Emission Sources*. EPA-450/3-81-013. September 1981.

35. U.S. EPA. *Source Category Survey: Secondary Zinc Smelting and Refinery Industry*. EPA-450/3-80-012. May 1980.

36. U.S. EPA. *Air Oxidation Processes in Synthetic Organic Chemical Manufacturing Industry - Background Information for Proposed Standards*. EPA-450/3-82-001a. October 1983.

37. U.S. EPA. *Reactor Processes in Synthetic Organic Chemical Manufacturing - Background Information for Proposed Standards* (Draft). October 1984.

38. U.S. EPA. *VOC Emissions from Volatile Organic Liquid Storage Tanks — Background Information for Proposed Standards* (Draft). EPA-450/3-81-003. July 1984.

39. U.S. EPA. *VOC Fugitive Emissions in Synthetic Organic Chemicals Manufacturing Industry - Background Information for Promulgated Standards*. EPA-450/3-80-033b. June 1982.

40. U.S. EPA. *Distillation Operations in Synthetic Organic Chemical Manufacturing - Background Information for Proposed Standards*. EPA-450/3-83-005a. December 1983.

41. U.S. EPA. *Organic Chemical Manufacturing Volume 6: Selected Processes*. EPA-450/3-80-028a. December 1980.

42. U.S. EPA. *Source Category: Ammonia Manufacturing Industry*. EPA-450/3-80-014. August 1980.

43. U.S. EPA. *Source Assessment: Ammonium Nitrate Production*. EPA-600/2-77-107i. September 1977.

44. U.S. EPA. *Ammonium Sulfate Manufacture - Background Information for Proposed Standards*. EPA-450/3-79-034a. December 1979.

45. U.S. EPA. *Preliminary Study of Sources of Inorganic Arsenic*. EPA-450/5-82-005. August 1982.

46. U.S. EPA. *Source Assessment: Major Barium Chemicals*. EPA-600/2-78-004b. March 1978.

47. U.S. EPA. *Emission Factors for Trace Substances*. EPA-450/2-73-001. December 1973.

48. U.S. EPA. *Review of New Source Performance Standards for Nitric Acid Plants*. EPA-450/3-84-011. April 1984.

49. U.S. EPA. *Sodium Carbonate Industry - Background Information for Proposed Standards*. EPA-450/3-80-029a. August 1980.

50. U.S. EPA. *Industrial Process Profiles for Environmental Use: Sulfur, Sulfur Oxides and Sulfuric Acid*. EPA-600/2-77-023w. February 1977.

51. U.S. EPA. *Final Guideline Document: Control of Sulfuric Acid Mist Emissions from Sulfuric Acid Production Plants*. EPA-450/2-77-019. September 1977.

52. U.S. EPA. *Source Assessment: Charcoal Manufacturing*. EPA-600/2-78-004z. December 1978.

53. U.S. EPA. *Locating and Estimating Air Emissions from Sources of Formaldehyde*. EPA-450/4-84-007e. March 1984.

54. U.S. EPA. *Locating and Estimating Air Emissions from Sources of Chloroform*. EPA-450/4-84-007c. March 1984.

55. U.S. EPA. *Locating and Estimating Air Emissions from Sources of Carbon Tetrachloride*. EPA-450/4-84-007b. March 1984.

56. U.S. EPA. *Locating and Estimating Air Emissions from Sources of Chlorobenzenes* (Draft). September 1984.

57. U.S. EPA. *Plastics and Resins Industry - Industrial Process Profiles for Environmental Use*. EPA-600/2-77-023j. February 1977.

58. U.S. EPA. *Locating and Estimating Air Emissions from Sources of Phosgene* (Draft). September 1984.

59. U.S. EPA. *Locating and Estimating Air Emissions from Sources of Acrylonitrile*. EPA-450/4-84-007a. March 1984.

60. U.S. EPA. *Locating and Estimating Air Emissions from Sources of Ethylene Dichloride*. EPA-450/4-84-007d. March 1984.

61. U.S. EPA. *Asphalt Roofing Manufacturing Industry - Background Information for Proposed Standards* (Draft). EPA-450/3-80-021a. June 1980.

62. U.S. EPA. *Trace Pollutant Emissions from the Processing of Nonmetallic Ores*. EPA-650/2-74-122. November 1974.

63. U.S. EPA. *Source Category Survey: Refractory Industry*. EPA-450/3-80-006. March 1980.

64. U.S. EPA. *A Review of Standards of Performance for New Stationary Sources - Portland Cement Industry*. EPA-450/3-79-012. March 1979.

65. U.S. EPA. *Background Information for Standards of Performance: Coal Preparation Plants Volume I: Proposed Standards*. EPA-450/2-74-021a. October 1974.

66. U.S. EPA. *Glass Manufacturing Plants, Background Information: Proposed Standards of Performance* (Draft). EPA-450/3-79-005a. June 1979.

67. U.S. EPA. *Wool Fiberglass Insulation Manufacturing Industry - Background Information for Proposed Standards* (Draft). EPA-450/3-83-002A. December 1983.

68. U.S. EPA. *Standards Support and Environmental Impact Statement Volume I: Proposed Standards of Performance for Lime Manufacturing Plants*. EPA-450/2-77-007a. April 1977.

69. U.S. EPA. *Final Standards Support and Environmental Impact Statement Volume II: Promulgated Standards of Performance for Lime Manufacturing Plants*. EPA-450/2-77-007b. October 1977.

70. U.S. EPA. *Source Category Survey: Mineral Wool Manufacturing Industry*. EPA-450/3-80-016. March 1980.

71. U.S. EPA. *Source Category Survey: Perlite Industry*. EPA-450/3-80-005. May 1980.

72. U.S. EPA. *Radionuclides - Background Information Document for Final Rules. Volume I*. EPA-520/1-84-022-1. October 1984.

73. *Standards Support and Environmental Impact Statement for the Iron Ore Benefication Industry* (Draft). Battelle Columbus Laboratories. December 1976.

74. U.S. EPA. *Kraft Pulping - Control of TRS Emissions from Existing Mills*. EPA-450/2-78-003b. March 1979.

75. U.S. EPA. *Industrial Process Profiles for Environmental Use: Chapter 2. Oil and Gas Production Industry*. EPA-600/2-77-023b. February 1977.

76. U.S. EPA. *Industrial Process Profiles for Environmental Use: Chapter 3. Petroleum Refining Industry*. EPA-600/2-77-023c. January 1977.

77. U.S. EPA. *Industrial Process Profiles for Environmental Use: Chapter 5. Basic Petrochemicals Industry*. EPA-600/2-77-023e. January 1977.

78. U.S. EPA. *VOC Fugitive Emissions in Petroleum Refining Industry - Background Information for Proposed Standards*. EPA-450/3-81-015a. November 1982.

79. U.S. EPA. *VOC Species Data Manual, Second Edition*. EPA-450/4-80-115. July 1980.

80. U.S. EPA. *Bulk Gasoline Terminals - Background Information for Proposed Standards* (Draft). EPA-450/3-80-038a. December 1980.

81. U.S. EPA. *Air Toxics Emission Patterns and Trends - Final Report*. EPA Contract No. 68-02-3513, Task 46. July 1984.

82. U.S. EPA. *Hazardous Emission Characterization of Utility Boilers*. EPA-600/2-75-066. July 1975.

83. U.S. EPA. *Thermal Conversion of Municipal Wastewater Sludge Phase II: Study of Heavy Metal Emissions*. EPA-600/2-81-203. September 1981.

84. U.S. EPA. *Locating and Estimating Air Emissions from Sources of Polychlorinated Biphenyls* (Draft). November 1984.

Chapter 3
Control Device Selection

3.1 Background

This chapter presents guidelines that will enable the user to select the control technique(s) that can be used to control HAP's. The control techniques that can be applied to control HAP emissions from a specific emission source will depend on the emission source characteristics and HAP characteristics. Therefore, Section 3.2, Vapor Emissions Control, and Section 3.3, Particulate Emissions Control, each pertain to specific HAP groups. The discussion of control technique selection within each section is according to type of HAP (organic or inorganic) and emission source (point, process fugitive, or area fugitive).

In the following sections, guidelines for selecting controls for point sources are discussed in detail. Point sources are typically controlled by add-on control devices. For each control technique, ranges of applicability with respect to emission stream characteristics, HAP characteristics, performance levels (e.g., removal efficiency), and other considerations that are important in control device selection are described in detail.

Work practices, including equipment modifications, play a key role in reducing emissions from process fugitive and area fugitive sources. These sources can also be controlled by add-on control devices if the emissions can be captured by hooding or enclosure or collected by closed vent systems and then transferred to a control device. Note that the overall performance of the control system will then be dependent on both the capture efficiency of the fugitive emissions and the efficiency of the control device.

To illustrate the control device selection process, several emission stream scenarios are presented throughout this chapter. The emission stream from the paper coating drying oven introduced in Chapter 2 is one of the scenarios presented. The data necessary for control device selection are recorded on the HAP Emission Stream Data Form (see Figure 2-1).

3.2 Vapor Emissions Control

3.2.1 Control Techniques for Organic Vapor Emissions from Point Sources

The most frequent approach to point source control is the application of add-on control devices. These devices can be of two types: combustion and recovery. Applicable combustion devices are thermal incinerators, catalytic incinerators, flares, and boilers/process heaters. Applicable recovery devices include condensers, adsorbers, and absorbers. The combustion devices are the more commonly applied control devices, since they are capable of high removal (i.e., destruction) efficiencies for almost any type of HAP (organic vapor). The removal efficiencies of the recovery techniques generally depend on the physical and chemical characteristics of the HAP under consideration.

Applicability of the control techniques depends more on the individual emission stream under consideration than on the particular source category (e.g., degreasing vs. surface coating in solvent usage operations source category). Thus, selection of applicable control techniques for point source emissions is made on the basis of stream-specific characteristics and desired control efficiency. Table 3-1 identifies the key emission stream characteristics and HAP characteristics that affect the applicability of each control technique and presents limiting values for each of these characteristics.

Matching the specific characteristics of the stream under consideration with the corresponding values in Table 3-1 will help the user to identify those techniques that can potentially be used to control the emission stream. The list of potentially applicable control techniques will then be narrowed further depending on the capability of the applicable control devices to achieve the required performance levels. Figure 3-1 identifies the expected emission reduction from the application of each control technique on the basis of the total VOC concentration in the emission stream. Very little data regarding control device removal efficiency

Table 3-1. Key Emission Stream and HAP Characteristics for Selecting Control Techniques for Organic Vapors from Point Sources

	Emission Stream Characteristics					HAP Characteristics[a]			
Control Device	HAP/Organics Contents[b] (ppmv)	Heat Content (Btu/scf)	Moisture Content (%)	Flow Rate (scfm)	Temp. (°F)	Molecular Weight (lb/lb-mole)	Solubility	Vapor Pressure (mm Hg)	Adsorptive Properties
Thermal Incinerator	>20; (<25% of LEL[c])			<100,000[d]					
Catalytic Incinerator	50-10,000; (<25% of LEL[c])			<100,000					
Flare		>300[e]		<2,000,000[f]					
Boiler/ Process Heater[g]		>150[h]		Steady					
Carbon Adsorber	1,000-10,000 (<25% of LEL[c])		50%[i]	300-100,000	100-200	45-130			Must be able to adsorb on/ desorb from available adsorbents
Absorber	250-10,000			1,000-100,000			Must be readily soluble in water or other solvents		
Condenser	>5,000			<2,000				>10 (at room temperature)	

[a] Refers to the characteristics of the individual HAP if a single HAP is present and to that of the HAP mixture if a mixture of HAP's is present.
[b] Determined from HAP/hydrocarbon content.
[c] For emission streams that are mixtures of air and VOC; in some cases, the LEL can be increased to 40 to 50% with proper monitoring and control (see Section 4.2 for definition of LEL).
[d] For packaged units; multiple-package or custom-made units can handle larger flows.
[e] Based on EPA's guidelines for 98% destruction efficiency.
[f] Units: lb/hr. Source: Reference 14.
[g] Applicable if such a unit is already available on site.
[h] Total heat content.
[i] Relative humidity.

Figure 3-1. Percent reduction ranges for add-on control devices.

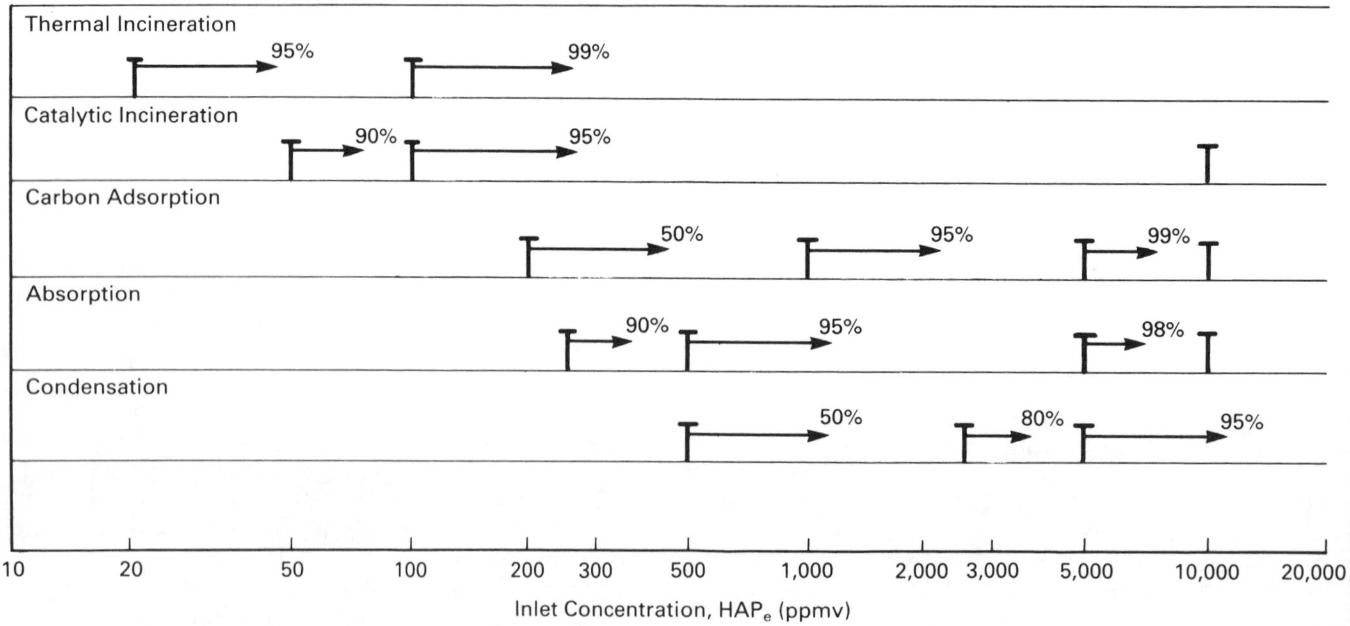

for specific HAP's are available. Therefore, without actual source test data for a specific emission stream and control system, HAP removal efficiency is assumed to equal total volatile organic compound (VOC) removal efficiency.

3.2.1.1 Thermal Incinerators

Thermal incinerators are used to control a wide variety of continuous emission streams containing VOC's. Compared to the other techniques, thermal incineration is broadly applicable; that is, it is much less dependent on HAP characteristics and emission stream characteristics. Destruction efficiencies up to 99+ percent are achievable with thermal incineration. Although they accommodate minor fluctuations in flow, thermal incinerators are not well suited to streams with highly variable flow because the reduced residence time and poor mixing during increased flow conditions decrease the completeness of combustion. This causes the combustion chamber temperature to fall, thus decreasing the destruction efficiency.

Thermal incineration is typically applied to emission streams that are dilute mixtures of VOC and air. In such cases, due to safety considerations, concentration of the VOC's is generally limited by insurance companies to 25 percent of the LEL (lower explosive limit) for the VOC in question (see Section 4.2.2 for more details). Thus, if the VOC concentration is high, dilution may be required.

When emission streams treated by thermal incineration are dilute (i.e., low heat content), supplementary fuel is required to maintain the desired combustion temperatures. Supplementary fuel requirements may be reduced by recovering the energy contained in the hot flue gases from the incinerator. For emission streams with high heat contents (e.g., > 150 Btu/scf), the possibility of using the emission stream as fuel gas should be considered.

Packaged single unit thermal incinerators are available in many sizes to control emission streams with flow rates from a few hundred up to about 100,000 scfm.

3.2.1.2 Catalytic Incinerators

Catalytic incinerators are similar to thermal incinerators in design and operation except that the former employ a catalyst to enhance the reaction rate. Since the catalyst allows the reaction to take place at lower temperatures, significant fuel savings are possible with catalytic incineration.

Catalytic incineration is not as broadly applied as thermal incineration since performance of catalytic incinerators is more sensitive to pollutant characteristics and process conditions than is thermal incinerator performance. Materials such as phosphorus, bismuth, lead, arsenic, antimony, mercury, iron oxide, tin, zinc, sulfur, and halogens in the emission stream can poison the catalyst and severely affect its performance. (Note: Some catalysts can handle emission streams containing halogenated compounds.) Liquid or solid particles that deposit on the catalyst and form a coating also reduce the catalyst's activity by preventing contact between the VOC's and the catalyst surface. Catalyst life is limited by thermal aging and by loss of active sites by erosion, attrition, and vaporization. With proper operating temperatures and adequate temperature control, these processes are normally slow, and satisfactory performance can be maintained for 3 to 5 years before replacement of the catalyst is necessary.

Catalytic incineration is generally less expensive than thermal incineration in treating emission streams with low VOC concentrations. Emission streams with high VOC concentrations should not be treated by catalytic incineration without dilution since such streams may cause the catalyst bed to overheat and lose its activity. Also, fluctuations in the VOC content of the emission stream should be kept to a minimum to prevent damage to the catalyst.

Destruction efficiencies of up to 95 percent of HAP's are generally achieved with catalytic incineration. Higher destruction efficiencies (99 percent) are also achievable, but require larger catalyst volumes and/or higher temperatures.

Catalytic incinerators have been applied to continuous emission streams with flow rates up to about 100,000 scfm.

3.2.1.3 Flares

Flares are commonly used for disposal of waste gases during process upsets (e.g., start-up, shutdown) and emergencies. They are basically safety devices that are also used to destroy waste emission streams.

Flares can be used for controlling almost any VOC emission stream. They can be designed and operated to handle fluctuations in emission VOC content, inerts content, and flow rate. There are several different types of flares including steam-assisted, air-assisted, and pressure head flares. Steam-assisted flares are very common and typically employed in cases where large volumes of waste gases are released. Air-assisted flares are generally used for moderate relief gas flows. Pressure head flares are small; they are used in arrays of up to 100 individual flares. Normally, only a few of the flares operate. The number of flares operating is increased as the gas flow increases.

Flaring is generally considered a control option when the heating value of the emission stream cannot be recovered because of uncertain or intermittent flow as in process upsets or emergencies. If the waste gas to be flared does not have sufficient heating value to sustain combustion, auxiliary fuel may be required.

Based on studies conducted by EPA, 98 percent destruction efficiency can be achieved by steam-assisted flares when controlling emission streams with heat contents greater than 300 Btu/scf. Depending on the type of flare configuration (e.g., elevated or ground flares), the capacity of flares to treat waste gases can vary—up to about 100,000 lb/hr for ground flares and 2 million lb/hr or more for elevated flares. The capacity of an array of pressure head flares depends on the number of flares in the array.

3.2.1.4 Boilers/Process Heaters

Existing boilers or process heaters can be used to control emission streams containing organic compounds. These are currently used as control devices for emission streams from several industries (e.g., refinery operations, SOCMI reactor processes and distillation operations, etc.)

Typically, emission streams are controlled in boilers or process heaters and used as supplemental fuel only if they have sufficient heating value (greater than about 150 Btu/scf). In some instances, emission streams with high heat content may be the main fuel to the process heater or boiler (e.g., process off-gas from ethylbenzene/styrene manufacturing). Note that emission streams with low heat content can also be burned in boilers or process heaters when the flow rate of the emission stream is small compared to the flow rate of the fuel/air mixture.

When used as emission control devices, boilers or process heaters can provide destruction efficiencies of greater than 98 percent at small capital cost and little or no fuel cost. In addition, near complete recovery of the emission stream heat content is possible.

There are some limitations in the application of boilers or process heaters as emission control devices. Since these combustion devices are essential to the operation of a plant, only those emission streams that will not reduce their performance or reliability can be controlled using these devices. Variations in emission stream flow rate and/or heating value could adversely affect the performance of a boiler or process heater. By lowering furnace temperatures, emission streams with large flow rates and low heating values can cause incomplete combustion and reduce heat output. The performance and reliability of the process heater or boiler may also be affected by the presence of corrosive compounds in the emission stream; such streams are usually not destroyed in these devices.

3.2.1.5 Carbon Adsorbers

Carbon adsorption is commonly employed as a pollution control and/or a solvent recovery technique. It is applied to dilute mixtures of VOC and air. Removal efficiencies of 95 to 99 percent can be achieved using carbon adsorption. The maximum practical inlet concentration is usually about 10,000 ppmv. The inlet concentrations are typically limited by the adsorption capacity of the carbon bed or safety problems posed by high bed temperatures produced by heat of adsorption and presence of flammable vapors. Outlet concentrations of 50 to 100 ppmv can be routinely achieved with state-of-the-art systems; concentrations as low as 10 to 20 ppmv can be achieved with some compounds. In contrast to incineration methods whereby the VOC's are destroyed, carbon adsorption provides a favorable control alternative when the VOC's in the emission stream are valuable.

High molecular-weight compounds that are characterized by low volatility are strongly adsorbed on carbon. The affinity of carbon for these compounds makes it difficult to remove them during regeneration of the carbon bed. Hence, carbon adsorption is not applied to such compounds (i.e., boiling point above 400°F; molecular weight greater than about 130). Highly volatile materials (i.e., molecular weight less than about 45) do not adsorb readily on carbon; therefore, adsorption is not typically used for controlling emission streams containing such compounds.

Carbon adsorption is relatively sensitive to emission stream conditions. The presence of liquid or solid particles, high boiling organics, or polymerizable substances may require pretreatment procedures such as filtration. Dehumidification is necessary if the emission stream has a high humidity (relative humidity $>$ 50 percent) and cooling may be required if the emission stream temperature exceeds 120° - 130°F.

To prevent excessive bed temperatures resulting from the exothermic adsorption process and oxidation reactions in the bed, concentrations higher than 10,000 ppmv must frequently be reduced. This is usually done by condensation or dilution of the emission stream ahead of the adsorption step. Exothermic reactions may also occur if incompatible solvents are mixed in the bed, leading to polymerization. If flammable vapors are present, the VOC concentrations may be limited by insurance companies to less than 25 percent of the LEL. If proper controls and monitors are used, LEL levels up to 40 to 50 percent may be allowed.

Packaged carbon adsorption systems are available that can handle emission streams with flow rates from a few hundred to above 100,000 scfm.

3.2.1.6 Absorbers (Scrubbers)

Absorption is widely used as a raw material and/or a product recovery technique in separation and purification of gaseous streams containing high concentrations of VOC's. As an emission control technique, it is much more commonly employed for inorganic vapors (e.g., hydrogen sulfide, chlorides, etc.) than for organic vapors. Using absorption as the primary control technique for organic vapor HAP's is subject to several limitations and problems as discussed below.

The suitability of absorption for controlling organic vapor emissions is determined by several factors; most of these factors will depend on the specific HAP in question. For example, the most important factor is the availability of a suitable solvent. The pollutant in question should be readily soluble in the solvent for effective absorption rates and the spent solvent should be easily regenerated or disposed of in an environmentally acceptable manner.

Another factor that affects the suitability of absorption for organic vapor emissions control is the availability of vapor/liquid equilibrium data for the specific HAP/solvent system in question. Such data are necessary for design of absorber systems. For uncommon HAP's, these data are not readily available.

Another consideration involved in the application of absorption as a control technique is disposal of the absorber effluent (i.e., used solvent). If the absorber effluent containing the organic compounds is discharged to the sewer, pond, etc., the air pollution problem is merely being transformed into a water pollution problem. Hence, this question should be addressed (e.g., are there chemical/physical/biological means for treating the specific effluent under consideration?). In solvent recovery, used organic solvents are typically stripped (reverse of absorption) and recycled to the absorber for economic reasons. However, in HAP control applications, stripping requirements will often be very expensive because the residual organic concentrations in the solvent must be extremely low for it to be suitable for reuse. Also, if the VOC's in the effluent from the absorber have appreciable vapor pressure (e.g., $>$ 0.1 mm Hg), the possibility of VOC emissions to the atmosphere should be considered.

In organic vapor HAP control applications, low outlet concentrations will typically be required. Trying to meet such requirements with absorption alone will lead to impractically tall absorption towers, long contact times, and high liquid-gas ratios that may not be cost effective. Therefore, absorbers will generally be effective when they are used in combination with other control devices such as incinerators.

Removal efficiencies in excess of 99 percent can be achieved with absorption.

3.2.1.7 Condensers

Condensers are widely used as raw material and/or product recovery devices. They are frequently applied as preliminary air pollution control devices for removing VOC contaminants from emission streams prior to other control devices such as incinerators, adsorbers, or absorbers.

Condensers are also used by themselves for controlling emission streams containing high VOC concentrations (usually $>$ 5,000 ppmv). In these cases, removal efficiencies obtained by condensers range from 50 to 90 percent. The removal efficiency of a condenser is dependent on the emission stream characteristics including the nature of the HAP in question (vapor pressure-temperature relationship) and HAP concentration, and the type of coolant used. Note that a condenser cannot lower the inlet VOC concentration to levels below the saturation concentration (or vapor pressure) at the coolant temperature. When water, the most commonly used coolant, is employed, the saturation conditions represent high outlet concentrations. For example, condenser outlet VOC concentrations are often limited to above 10,000 to 20,000 ppmv due to the saturation conditions of most of the organic compounds at the temperature of the cooling water. Therefore, it is not possible for condensation with water as the coolant to achieve the low outlet concentrations that would be required in HAP control applications.

Removal efficiencies above 90 percent can be achieved if lower temperatures than those possible with cooling water are employed. These low temperatures can be obtained with coolants such as chilled water, brine solutions, or chlorofluorocarbons. However, for extremely low outlet HAP concentrations, condensation will usually be economically infeasible.

Depending on the type of condenser used, there may be potential problems associated with the disposal of the spent coolant. Therefore, using contact condensers that generate such effluents for controlling HAP emissions is not recommended.

Flow rates up to about 2,000 scfm can be considered as representative of the typical range for condensers used as emission control devices. Condensers for emission streams with flow rates above 2,000 scfm and containing high concentrations of noncondensibles will require prohibitively large heat transfer areas.

3.2.2 Control Techniques for Inorganic Vapor Emissions From Point Sources

Inorganic vapors make up only a small portion of the total HAP's emitted to the atmosphere. Potential sources of the various inorganic vapors found in the atmosphere are discussed in Chapter 2. Inorganic HAP vapors typically include gases such as ammonia, hydrogen sulfide, carbonyl sulfide, carbon disulfide, metals with hydride and carbonyl complexes, chlorides, oxychlorides, and cyanides.

In many cases, although the inorganic HAP's are emitted as vapors at the emission source, they may condense when passing through various ducts and form particulates. Prior to discharge to the atmosphere, these particulates are typically controlled by methods that will be discussed in Section 3.3. In this section, the discussion will be based on control techniques for HAP's that are emitted as vapors to the atmosphere.

Only a limited number of control methods are applicable to inorganic vapor emissions from point sources. The two most commonly used control methods are absorption (scrubbing) and adsorption. Absorption is the most widely used and accepted method for inorganic vapor control. Although combustion can be used for some inorganic HAP's (e.g., hydrogen sulfide, carbonyl sulfide, nickel carbonyl), typical combustion methods such as thermal and catalytic incineration are generally not applied. In some cases, for example, in controlling hydrogen sulfide emissions from gas wells and gas processing, flares are used.

Applicability of absorption and adsorption as control methods depends on the individual emission stream characteristics. The removal efficiencies that can be achieved will be determined by the physical and chemical properties of the HAP under consideration. Other factors (e.g., waste disposal, auxiliary equipment requirements), while not necessarily affecting the technical feasibility of the control device, may affect the decision to use that particular control method.

In the following two subsections, the applicability of absorption and adsorption for controlling inorganic vapor emissions will be discussed.

3.2.2.1 Absorbers (Scrubbers)

Absorption is the most widely used recovery technique for separation and purification of inorganic vapor emissions. The removal efficiency achievable with absorbers can be greater than 99 percent. It will typically be determined by the actual concentrations of the specific HAP in gas and liquid streams and the corresponding equilibrium concentrations. Table 3-2 summarizes the reported efficiencies for various inorganic vapors employing absorption as the control method.

As discussed in Section 3.2.1.6 for organic vapors, the suitability of absorption for controlling inorganic vapors in gaseous emission streams is dependent on several factors. The most important factor is the solubility of the pollutant vapor in the solvent. The ideal solvent should be nonvolatile, noncorrosive, nonflammable, nontoxic, chemically stable, readily available, and inexpensive. Typical solvents used by industry for inorganic vapor control include water, sodium hydroxide solutions, amyl alcohol, ethanolamine, weak acid solutions, and hypochlorite solutions. Other factors which may affect inorganic vapor absorption are similar to those for organic vapor absorption (see Section 3.2.1.6).

Water is the ideal solvent for inorganic vapor control by absorption. It offers distinct advantages over other solvents, the main one being its low cost. It is typically used on a once-through basis and then discharged to a wastewater treatment system. The effluent may require pH adjustment to precipitate metals and other HAP's as hydroxides or salts; these are typically less toxic and can be more easily disposed of.

3.2.2.2 Adsorbers

When the removal of inorganic vapors is especially difficult using absorption methods, adsorption may prove to be more effective. Adsorbents such as activated carbon, impregnated activated carbon, silica gel, and activated alumina are capable of adsorbing various inorganic vapors and gases. The degree of adsorption is dependent not only on the waste stream characteristics, but also on the different characteristics of the adsorbents.

Carbon adsorption, using conventional and chemically impregnated carbons, is widely used for controlling inorganic vapors such as mercury, nickel carbonyl, phosgene, and amines. For example, when mercury vapors are passed through a bed of sulfur-impregnated carbon, the mercury vapors react with the sulfur to form a stable mercuric sulfide. Over 95 percent of the mercury removed in this way can be recovered for reuse.

Important factors to consider when choosing an adsorbent for inorganic vapor control are very similar to those for organic vapor control, which are discussed in Section 3.2.1.5. Some of these factors include the amount of adsorbent needed, temperature rise of the gas stream due to adsorption, ease of regeneration, and the useful life of the adsorbent. Most of the reported removal efficiencies for inorganic vapors are for activated carbon and impregnated activated carbon, and range from 90 to 100 percent. Table 3-2 summarizes removal efficiencies reported for various inorganic vapors controlled by adsorption.

Activated carbons are the most widely used adsorbents for inorganic vapor control. In several cases, they must be treated (i.e., impregnated with chemicals) for effective application. Since activated carbons are relatively sensitive to emission stream conditions, pretreatment of the emission stream may be necessary. Pretreatment methods such as filtration, cooling, and dehumidification may be required depending on the emission stream conditions. Filtration is used to prevent plugging of the adsorber bed by any solids or particles which may be in the emission stream. Ideal adsorption conditions for impregnated activated carbons are relative humidities less than 50 percent and gas stream temperatures below 130°F. Inorganic vapor concentrations are not recommended to exceed 1,000 ppmv (preferably, less than 500 ppmv) when activated carbon is used as an adsorbent.

3.2.3 Control Techniques for Organic/Inorganic Vapor Emissions From Process Fugitive Sources

Process fugitive emissions are defined in this handbook as emissions from a process or piece of equipment that are being emitted at locations other than the main vent or process stack. Process fugitive emissions include fumes or gases which escape from or through access ports and feed and/or discharge openings to a process. Examples include the open top of a vapor degreaser, the slag or metal tap opening on a blast furnace, and the feed chute on a ball mill. Process fugitive emission sources can also include vent fans from rooms or enclosures containing an emissions source. An example would be a vent fan on a perchloroethylene dry cleaner or the vent fan on a press room. Other examples of process fugitive sources include cooling towers and process drains.

These sources can be controlled by add-on control devices once the emissions from the sources are captured by hooding, enclosures, or closed vent systems and then transferred to a control device. Because of the nature of the opening (e.g., for access or maintenance), the opening through which emissions escape cannot be totally enclosed or blanked off. Operators have to access the equipment or materials have to be fed or discharged from the process. For this reason, hoods or partial enclosures are used to control emissions from such openings.

Proper hood design requires a sufficient knowledge of the process or operation so that the most effective hood or enclosure can be installed to provide minimum exhaust volumes for effective contaminant control. In theory, hood design is based upon trying to enclose the process and keep all openings to a minimum and located away from the natural path of containment travel. Where possible, inspection and maintenance openings should be provided with doors. In practice, hoods are designed using the capture velocity principle which involves creation of an air flow past the source of containment sufficient to remove the highly contaminated air from around the source or issuing from that source and draw the air into an exhaust hood. The capture velocity principle is based on the fact that small dust particles travel very short distances (on the order of inches) when thrown or emitted from a source and therefore can be assumed to follow air currents. Vapors and gases exhibit the same effects.

In practice, hood capture efficiency is very difficult to determine and therefore, when evaluating a hood, one of the few parameters that can be considered is the capture velocity. Standard design values of capture velocity are available from the American Conference of Government Industrial Hy-

Table 3-2. Current Control Methods for Various Inorganic Vapors (1)

	Absorption		Adsorption	
Inorganic Vapor	Reported Removal Efficiency (%)	Solvent	Reported Removal Efficiency (%)	Adsorbent
Mercury (Hg)	95	Brine/Hypochlorite solution	90	Sulfur-impregnated activated carbon
Hydrogen Chloride (HCl)	95	Water		
Hydrogen Sulfide (H_2S)	98	Sodium carbonate/Water	100	Ammonia-impregnated activated carbon
Calcium Fluoride (CaF_2)	95	Water		
Silicon Tetrafluoride (SiF_4)	95	Water		
Hydrogen Fluoride (HF)	85-95	Water	99	Calcined alumina
Hydrogen Bromide (HBr)	99.95	Water		
Titanium Tetrachloride ($TiCl_4$)	99	Water		
Chlorine (Cl_2)	90	Alkali solution		
Hydrogen Cyanide (HCN)				Ammonia-impregnated activated carbon

gienists in the Industrial Ventilation Manual (see Table 3-3).(2) Once a capture velocity has been determined, the volume of air required should be based on maintaining this capture velocity at the emissions point furthest from the hood. This capture velocity should be sufficient to overcome any opposing air currents. (For additional information on hood design guidelines for several industries, see Appendix A.7, reference 3, or reference 2.)

Very few measurements of hood capture efficiency have been conducted.(4,5,6) Hood capture efficiencies of between 90 and 100 percent are possible depending on the situation and the particular process fugitive sources being controlled. For sources where operator access is not needed and where inspection doors can be provided, efficiencies toward the upper end of the range are achievable. For sources where emissions are more diffused, for example, from printing presses, capture efficiencies of 90 percent may be difficult. In the flexible vinyl and printing industry, 90 percent is typically the upper bound for capture efficiency on coating presses.(4) In the publication rotogravure industry, capture efficiencies of 93 to 97 percent have been demonstrated based on material balances.(5)

Once the process fugitive emissions are captured, the selection of the control device will be dependent on the emission stream characteristics, HAP characteristics, and the required overall performance levels (e.g., removal efficiency). Note that the required performance level for the control device will be determined by the capture efficiency. The factors that affect the control device selection process are the same as for point sources; therefore, refer to Sections 3.2.1 and 3.2.2.

For process fugitive emission sources such as process drains, the control alternatives involve a closure or a seal. A common method involves the use of a P-leg in the drain line with a water seal. A less common, but more effective method, is a completely closed drain system. Several factors affect the performance of water-sealed drains in reducing organic emissions: drainage rate, composition and temperature of the liquid entering the drain, diameter of the drain, and ambient atmospheric conditions. Emission reductions of 40 to 50 percent may be achieved with water-sealed drains.(7) In a completely closed drain system, the system may be pressured and purged to a control device to effectively capture all emissions. The control efficiency would then depend on the efficiency of the control device; 95 percent should be achievable.(7)

Process fugitive emissions from cooling towers have not been reliably quantified due to difficulties encountered in measuring them and no specific control guidelines have been developed.(8) Probably the best control technique currently available is close monitoring of heat exchangers and other equipment to detect small leaks as they occur.

3.2.4 Control Techniques for Organic/Inorganic Vapor Emissions from Area Fugitive Sources

The control measures that can be employed for controlling organic or inorganic vapor emissions from area fugitive sources are basically the same. Organic fugitive emissions have been more extensively studied than inorganic fugitive emissions. Therefore, the following discussion will be primarily based on organic vapors. However, control techniques for inorganic vapor emissions will also be discussed.

Table 3-3. Range of Capture Velocities (2)

Condition of Contaminant Dispersion	Examples	Capture Velocity (fpm[a])
Released with practically no velocity into quiet air	Evaporation from tanks; degreasing, etc.	50-100
Released at low velocity into moderately still air	Spray booths; intermittent container filling; low speed conveyor transfers; welding; plating; pickling	100-200
Active generation into zone of rapid air motion	Spray painting in shallow booths; barrel filling; booths; barrel filling; conveyor loading; crushers	200-500
Released at high initial velocity into zone of very rapid air motion	Grinding; abrasive blasting; tumbling	500-2,000

In each category above, a range of capture velocity is shown. The proper choice of values depends on several factors:

Lower End of Range:
- Room air currents minimal or favorable to capture.
- Contaminants of low toxicity or of nuisance value only.
- Intermittent, low production.
- Large hood-large air mass in motion.

Upper End of Range:
- Disturbing room air currents.
- Contaminants of high toxicity.
- High production, heavy use.
- Small hood-local control only.

[a] fpm = feet per minute.

Fugitive emissions of organic vapors occur in plants processing organic liquids and gases, such as petroleum refineries, chemical plants, and plants producing chemically based products such as plastics, dyes, and drugs. One group of emission sources found in plants of this type is commonly referred to as equipment leaks. Fugitive emissions of this type result from incomplete sealing of equipment at the point of interface of process fluid with the environment. Control techniques for equipment leaks include leak detection and repair programs and equipment installation or configuration. The following sections contain information about control techniques for common types of processing equipment found in plants processing organic materials. Control techniques and control efficiencies for common types of processing equipment are summarized in Table 3-4.

Table 3-4. Summary of Control Effectiveness for Controlling Organic Area Fugitive Emission Sources (9)

Emission Source	Control Technique Equipment Modification	Control Effectiveness (8) (%)
Pumps	Monthly leak detection and repair	61
	Sealless pumps	100
	Dual mechanical seals	100
	Closed vent system[a]	100
Valves		
- Gas	Monthly leak detection and repair	73
	Diaphragm valves	100
- Light liquid	Monthly leak detection and repair	46
	Diaphragm valves	100
Pressure Relief Valves	Rupture disk	100
	Closed vent system[a]	100
Open-ended Lines	Caps, plugs, blinds	100
Compressors	Mechanical seals with vented degassing reservoirs	100
	Closed vent system[a]	100
Sampling Connections	Closed purge sampling	100

[a]Closed vent systems are used to collect and transfer the fugitive emissions to add-on control devices such as flares, incinerators, or vapor recovery systems.

Pumps—Several types of equipment or equipment configurations can be used to eliminate or capture all organic vapors leaking from pump seals. There are, first of all, leakless pumps, pumps designed with no interface between the process fluid and the environment, such as diaphragm seal, and canned pumps. These pumps effectively limit fugitive emissions. However, they are limited in application.

Sophisticated pump seals can also be used to capture or eliminate fugitive emissions. Dual seal systems with pressurized barrier fluids or low pressure systems vented to control devices may be used in some applications. Another approach which may be used involves venting the entire seal area to a control device. Capture efficiencies should be virtually 100 percent in both systems vented to control devices. Then the overall control efficiency would be limited by the efficiency of the control device.

Another approach to reducing (but not eliminating) organic fugitive emissions from pumps is leak detection and repair programs. A leak detection and repair system modeled after the one EPA developed for the New Source Performance Standards (NSPS) of the synthetic organic chemical industry (SOCMI) should achieve about 60 percent control efficiency.(9) The efficiency of leak detection and repair programs is dependent on several factors such as frequency of monitoring, effectiveness of maintenance, action level, and underlying tendency to leak. These factors and their effect on control efficiency have been studied and discussed in references 9, 10, 11 and 12. Also, models are available for calculating the effectiveness of leak detection and repair programs (e.g., see reference 13).

Valves—As with pumps, control of fugitive emissions from valves may be accomplished by installing equipment designed to isolate the process fluid from the environment. But also as with pumps, leakless valves such as diaphragm valves are limited in their application.

Leak detection and repair programs have been used to reduce fugitive emissions from valves. As indicated above for pumps, the control efficiency of a leak detection and repair program depends on the frequency of monitoring, the effectiveness of maintenance, the action level, the underlying tendency to leak, and other factors. A leak detection and repair program modeled after the one developed by EPA for the NSPS of the SOCMI should achieve a control efficiency of about 70 percent for valves in gas service and about 50 percent for valves in light liquid service.(9)

Pressure Relief Valves—Fugitive emissions from pressure relief valves may be virtually eliminated through the use of rupture disks to prevent leakage through the seal. Fugitive emissions may also be added to gases collected in a flare system by piping the relief valve to a flare header. The control efficiency, then, depends on the destruction efficiency of the flare. If flares are operated in accordance with flare requirements recently established by EPA for sources complying with NSPS, at least 98 percent control efficiency should be achieved.

Open-ended Lines—Leakage of organic vapors through valve seats to the open ends of pipes can be eliminated by the installation of caps, plugs, or blind flanges. The control efficiency should be 100 percent as long as the plugs and caps remain in place.

Compressors—Fugitive emissions from compressor seals can be controlled by venting the seal area to a flare or other control device. Barrier fluid systems can also be used to purge the seal area and convey leakage to a control device. Capture efficiency should be 100 percent and, therefore, the overall control efficiency would depend on the efficiency of the control device.

Sampling Connections—Fugitive emissions from sampling connections can be controlled by returning the purged material to the process or by disposing of it in a control device. The practice of returning purged material to the process in a closed system should achieve almost 100 percent control efficiency. The control efficiency achieved by diverting the collected purge material to a control device depends on the efficiency of the device.

Other area fugitive emission sources include lagoons and ponds where liquid waste streams containing organic compounds are disposed of. Emissions and emission rates of organic vapors from such sources are not well documented, and such sources are not easily controlled. The best method currently available for reducing emissions from lagoons and ponds is enhancing upstream treatment processes, thereby minimizing the amount of organic material reaching the lagoons and/or ponds.

Area fugitive emissions of inorganic vapors may be found in plants processing inorganic chemicals, metals, electronics, and other products. Although extensive work has not been done to quantify equipment leaks in plants processing inorganic chemicals, it is expected that they would be similar to equipment leaks encountered in plants processing organic chemicals. Therefore, plants processing highly volatile compounds such as hydrogen chloride or ammonia would be expected to benefit by the same control techniques applied to reduce or eliminate fugitive emissions containing organic compounds. There are a couple of differences to keep in mind, however. First, for those control techniques that employ a control device to treat collected vapors, the control device will probably differ. Instead of a combustion device, an absorber, condenser, or an adsorber may be a more appropriate choice. Another difference would be relevant to leak detection and repair programs. Leak detection and repair programs for organic vapors were developed using portable organic analyzers. Portable analyzers that respond to the inorganic vapors of concern would have to be used for leak detection of inorganic materials.

Controlling inorganic vapor emissions from area sources such as lagoons and/or ponds where liquid waste streams containing volatile inorganic compounds are disposed of is quite difficult. The best control method currently available is minimizing the quantity of inorganic compounds reaching the lagoon and/or pond by improving upstream treatment processes.

3.2.5 Control Device Selection for a Hypothetical Facility

This subsection illustrates the control device selection process discussed in the previous sections for a hypothetical facility with several emission streams. Assume that the owner/operator of this facility has requested assistance regarding the control of these emission streams. The data supplied by the owner/operator are presented in Figures 3-2 through 3-7.

Emission Stream 1 (see Figure 3-2)—This stream is the same as that described in Chapter 2 in the Example Case. Assuming the HAP control requirement is 99 percent reduction, from Figure 3-1, the only applicable control technique for this level of performance at concentration levels of ~100 ppmv is thermal incineration. The HAP concentration is less than 25 percent of the LEL for the HAP (see Table B.1-1); hence, the concentration limit indicated in Table 3-1 will not be exceeded. Also, the flow rate of Emission Stream 1 falls in the range of application indicated for thermal incinerators in Table 3-1.

Emission Stream 2 (see Figure 3-3)—Assume the HAP control requirement for Emission Stream 2 is 95 percent reduction. For this level of performance, the applicable control techniques for inlet concentrations of ~500 ppmv are thermal incineration, catalytic incineration, and absorption. If either of the incineration techniques are applied, the concentration limit indicated in Table 3-1 will not be exceeded since the HAP concentration is less than 25 percent of the LEL for the HAP (see Table B.1-1). The flow rate of Emission Stream 2 falls in the range indicated as applicable in Table 3-1 for these control techniques. The final selection of the control technique should also be based on design criteria (Chapter 4) and costs (Chapter 5).

Emission Stream 3 (see Figure 3-4)—Assume the HAP control requirement for Emission Stream 3 is 98 percent reduction. In this case, the inlet HAP concentration falls outside the range indicated in Figure 3-1; therefore, none of the control devices in this figure are applicable. According to Section 3.2.1.3, flares can be used to control emission streams with high heat contents; hence, flaring can be considered an option. Also, if a boiler or a process heater is available on site, it can be used to control Emission Stream 3.

Emission Stream 4 (see Figure 3-5)—Assume the HAP control requirement for Emission Stream 4 is 95 percent reduction. For this level of performance, the applicable control techniques for inlet concen-

Figure 3-2. Effluent characteristics for emission stream #1.

HAP EMISSION STREAM DATA FORM*

Company	Glaze Chemical Company	Plant Contact	Mr. John Leake
Location (Street)	87 Octane Drive	Telephone No.	(999) 555-5024
(City)	Somewhere	Agency Contact	Mr. Efrem Johnson
(State, Zip)		No. of Emission Streams Under Review	6

A. Emission Stream Number/Plant Identification #1 / #3 Oven Exhaust

		(a)	(b)	(c)
B.	HAP Emission Source	paper coating oven	-	-
C.	Source Classification	process point	-	-
D.	Emission Stream HAP's	toluene	-	-
E.	HAP Class and Form	organic vapor	-	-
F.	HAP Content (1,2,3)**	73 ppmv	-	-
G.	HAP Vapor Pressure (1,2)	28.4 mm Hg @ 77°F	-	-
H.	HAP Solubility (1,2)	insoluble in water	-	-
I.	HAP Adsorptive Prop. (1,2)	provided	-	-
J.	HAP Molecular Weight (1,2)	92 lb/lb-mole	-	-

K.	Moisture Content (1,2,3)	2% vol.	P.	Organic Content (1)***	44 ppmv CH_4, 4 ppmv ot
L.	Temperature (1,2,3)	120°F	Q.	Heat/O_2 Content (1)	0.4 Btu/scf / 20.6% vol.
M.	Flow Rate (1,2,3)	15,000 scfm (max)	R.	Particulate Content (3)	
N.	Pressure (1,2)	atmospheric	S.	Particle Mean Diam. (3)	
O.	Halogen/Metals (1,2)	none / none	T.	Drift Velocity/SO_3 (3)	/

U. Applicable Regulation(s)

V. Required Control Level

W. Selected Control Methods

*The data presented are for an emission stream (single or combined streams) prior to entry into the selected control method(s). Use extra forms if additional space is necessary (e.g., more than three HAP's). and note this need.

**The numbers in parentheses denote what data should be supplied depending on the data in steps "C" and "E":
1 = organic vapor process emission
2 = inorganic vapor process emission
3 = particulate process emission

***Organic emission stream combustibles less HAP combustibles shown on Lines D and F.

trations of ~1,000 ppmv are thermal incineration, catalytic incineration, carbon adsorption, and absorption. If either of the incineration methods or carbon adsorption is applied, the concentration limit indicated in Table 3-1 will not be exceeded since the HAP concentration is less than 25 percent of the LEL (see Table B.1-1). The flow rate of Emission Stream 4 falls in the range indicated as applicable in Table 3-1 for these control techniques. The final selection of the control technique should also be based on design criteria (Chapter 4) and costs (Chapter 5).

Emission Stream 5 (see Figure 3-6)—Assume the HAP control requirement for Emission Stream 5 is 98 percent reduction. Since this emission stream contains inorganic HAP's, incineration techniques are not applicable. The only control technique that is applicable for this level of performance and inlet HAP concentrations of ~20,000 ppmv is absorption. The flow rate of Emission Stream 5 falls in the range indicated as applicable for absorption.

Emission Stream 6 (see Figure 3-7)—Assume the HAP control requirement for Emission Stream 6 is 90 percent. Also assume that the owner/operator has indicated his preference to recover the HAP in the emission stream. For this level of performance, the applicable control techniques for inlet concentrations of ~13,000 ppmv are absorption and condensation. The final selection among these techniques should also be based on design criteria (Chapter 4) and costs (Chapter 5).

3.3 Particulate Emissions Control

Section 3.3.1 discusses add-on particulate control devices and presents guidelines that are used to determine the applicability of each control device. Section 3.3.2 discusses control techniques that are used to reduce fugitive particulate emissions.

3.3.1 Control Techniques for Particulate Emissions from Point Sources

Three types of control devices applicable to particulate-laden emission streams from point sources are discussed below: fabric filters (baghouses), electrostatic precipitators (ESP), and venturi scrubbers. The control efficiencies and applicability of these devices depend on the physical and chemi-

Figure 3-3. Effluent characteristics for emission stream #2.

HAP EMISSION STREAM DATA FORM*

Company	Glaze Chemical Company	Plant Contact	Mr. John Leake
Location (Street)	87 Octane Drive	Telephone No.	(999) 555-5024
(City)	Somewhere	Agency Contact	Mr. Efrem Johnson
(State, Zip)		No. of Emission Streams Under Review	6 /

A. Emission Stream Number/Plant Identification #2 / #1 Oven Exhaust

		(a)	(b)	(c)
B.	HAP Emission Source	metal coating oven	-	-
C.	Source Classification	process point	-	-
D.	Emission Stream HAP's	toluene	-	-
E.	HAP Class and Form	organic vapor	-	-
F.	HAP Content (1,2,3)**	500 ppmv	-	-
G.	HAP Vapor Pressure (1,2)	28.4 mm Hg @ 77°F	-	-
H.	HAP Solubility (1,2)	insoluble in water	-	-
I.	HAP Adsorptive Prop. (1,2)	provided	-	-
J.	HAP Molecular Weight (1,2)	92 lb/lb-mole	-	-

K. Moisture Content (1,2,3)	2% vol.	P. Organic Content (1)***	none
L. Temperature (1,2,3)	120°F	Q. Heat/O_2 Content (1)	2.1 Btu/scf / 20.6% vol.
M. Flow Rate (1,2,3)	20,000 scfm (max)	R. Particulate Content (3)	
N. Pressure (1,2)	atmospheric	S. Particle Mean Diam. (3)	
O. Halogen/Metals (1,2)	none / none	T. Drift Velocity/SO_3 (3)	/

U. Applicable Regulation(s)

V. Required Control Level assume 95% removal

W. Selected Control Methods thermal incineration, catalytic incineration, absorption

*The data presented are for an emission stream (single or combined streams) prior to entry into the selected control method(s). Use extra forms if additional space is necessary (e.g., more than three HAP's). and note this need.

**The numbers in parentheses denote what data should be supplied depending on the data in steps "C" and "E":
1 = organic vapor process emission
2 = inorganic vapor process emission
3 = particulate process emission

***Organic emission stream combustibles less HAP combustibles shown on Lines D and F.

cal/electrical properties of the airborne particulate matter under consideration. Brief descriptions of each of these control devices appear in the subsections that follow.

Selection of the control devices themselves depends on the specific stream characteristics and the parameters (e.g., required collection efficiency) that affect the applicability of each control device. Table 3-5 identifies some key emission stream characteristics that affect the applicability of each device. Matching the characteristics of the emission stream under consideration with the corresponding information presented in Table 3-5 will identify those techniques most suited to control the stream. This does not imply that a given control device should be excluded at this point, however. In general, the parameters listed in Table 3-5 are given as typical guidelines and should not be taken as absolute, definitive values. Gas stream pretreatment equipment can be installed upstream of the control device (i.e., cyclones, precoolers, preheaters) which enables the emission stream to fall within the parameters outlined in Table 3-5.

The temperature of the emission stream should be within 50 to 100°F above its dew point if the emission stream is to be treated (i.e., particulate matter collected) by an ESP or a fabric filter. If the emission stream temperature is below this range, condensation can occur; condensation can lead to corrosion of metal surfaces, blinding and/or deterioration of fabric filter bags, etc. If the emission stream is above this range, optimal HAP collection may not occur; by lowering the emission stream temperature, the vapor component of the HAP is reduced and, thus, an ESP or fabric filter will collect the HAP more effectively. Procedures for determining the dew point of an emission stream are provided in Appendix B.1. (For discussions of gas stream pretreatment equipment, see reference 15 and Appendix B.11 of reference 3.)

Table 3-6 identifies general advantages and disadvantages for each particular control device. Table 3-6 is used to provide additional information on other considerations that, while not necessarily affecting the technical feasibility of the control device for the stream, may affect the overall desirability of

Figure 3-4. Effluent characteristics for emission stream #3.

HAP EMISSION STREAM DATA FORM*

Company Glaze Chemical Company
Location (Street) 87 Octane Drive
(City) Somewhere
(State, Zip)

Plant Contact Mr. John Leake
Telephone No. (999) 555-5024
Agency Contact Mr. Efrem Johnson
No. of Emission Streams Under Review 6

		(a)	(b)	(c)
A.	Emission Stream Number/Plant Identification	#3 / Acetaldehyde Manufacturing Absorber Vent		
B.	HAP Emission Source	absorber vent	-	-
C.	Source Classification	process point	-	-
D.	Emission Stream HAP's	methylene chloride	-	-
E.	HAP Class and Form	organic vapor	-	-
F.	HAP Content (1,2,3)**	44,000 ppmv	-	-
G.	HAP Vapor Pressure (1,2)	436 mm Hg @ 77°F	-	-
H.	HAP Solubility (1,2)	insoluble in water	-	-
I.	HAP Adsorptive Prop. (1,2)	not given	-	-
J.	HAP Molecular Weight (1,2)	85 lb/lb-mole	-	-

K. Moisture Content (1,2,3) none
L. Temperature (1,2,3) 100°F
M. Flow Rate (1,2,3) 30,000 scfm (expected)
N. Pressure (1,2) atmospheric
O. Halogen/Metals (1,2) none / none

P. Organic Content (1)*** 17.8% vol. CH_4
Q. Heat/O_2 Content (1) 180 Btu/scf / none
R. Particulate Content (3)
S. Particle Mean Diam. (3)
T. Drift Velocity/SO_3 (3) /

U. Applicable Regulation(s)
V. Required Control Level assume 98% removal
W. Selected Control Methods flare, boiler, process heater

*The data presented are for an emission stream (single or combined streams) prior to entry into the selected control method(s). Use extra forms if additional space is necessary (e.g., more than three HAP's). and note this need.

**The numbers in parentheses denote what data should be supplied depending on the data in steps "C" and "E":
1 = organic vapor process emission
2 = inorganic vapor process emission
3 = particulate process emission

***Organic emission stream combustibles less HAP combustibles shown on Lines D and F.

Table 3-5. Key Characteristics for Particulate Emission Streams

Control Device	Achievable Efficiency Range	Particle Size Limitation	Temperature	Corrosiveness/ Resistivity	Moisture Content
Baghouse	Up to 99+%	Least efficient with particles 0.1 μm to 0.3 μm diameter.	Dependent on fiber type but not exceeding 550°F without a precooler	Special fiber types necessary to resist corrosion.	Poor efficiency with emission streams of high moisture content, very sensitive to changes in moisture content of an emission stream.
ESP	Up to 99+%	Generally least efficient with particles ranging in size from 0.2 μm to 0.5 μm diameter.	Generally up to 1,000°F	Corrosion resistant materials required. May require conditioning agents for highly resistive particles. Additionally, ESP's are not used to control organic matter since this constitutes a fire hazard.	Can control streams with relatively high moisture content (i.e., $34\%_{vol}$) if so designed, but sensitive to moisture changes of an emission stream.
Venturi	Up to 99+%	Generally operates best with particles >0.5 μm diameter.	No general limitations	Special construction may be required for corrosive emission streams.	Not sensitive to changes in moisture content of emissions stream.

Figure 3-5. Effluent characteristics for emission stream #4.

HAP EMISSION STREAM DATA FORM*

Company	Glaze Chemical Company	Plant Contact	Mr. John Leake
Location (Street)	87 Octane Drive	Telephone No.	(999) 555-5024
(City)	Somewhere	Agency Contact	Mr. Efrem Johnson
(State, Zip)		No. of Emission Streams Under Review	6

A. Emission Stream Number/Plant Identification #4 / #1 Printing Press Exhaust

		(a)	(b)	(c)
B.	HAP Emission Source	printing press	-	-
C.	Source Classification	process point	-	-
D.	Emission Stream HAP's	toluene	-	-
E.	HAP Class and Form	organic vapor	-	-
F.	HAP Content (1,2,3)**	1,000 ppmv	-	-
G.	HAP Vapor Pressure (1,2)	28.4 mm Hg @ 77°F	-	-
H.	HAP Solubility (1,2)	insoluble in water	-	-
I.	HAP Adsorptive Prop. (1,2)	provided	-	-
J.	HAP Molecular Weight (1,2)	92 lb/lb-mole	-	-

K.	Moisture Content (1,2,3)	40% rel. humidity	P.	Organic Content (1)***	none
L.	Temperature (1,2,3)	90°F	Q.	Heat/O_2 Content (1)	4.2 Btu/scf / 20.6% vol.
M.	Flow Rate (1,2,3)	15,000 scfm (max)	R.	Particulate Content (3)	
N.	Pressure (1,2)	atmospheric	S.	Particle Mean Diam. (3)	
O.	Halogen/Metals (1,2)	none / none	T.	Drift Velocity/SO_3 (3)	/

U. Applicable Regulation(s)

V. Required Control Level assume 95% removal

W. Selected Control Methods thermal incineration, catalytic incineration, carbon adsorption, absorption

*The data presented are for an emission stream (single or combined streams) prior to entry into the selected control method(s). Use extra forms if additional space is necessary (e.g., more than three HAP's). and note this need.

**The numbers in parentheses denote what data should be supplied depending on the data in steps "C" and "E":
1 = organic vapor process emission
2 = inorganic vapor process emission
3 = particulate process emission

***Organic emission stream combustibles less HAP combustibles shown on Lines D and F.

using the device for a given emission stream. Thus, Tables 3-5 and 3-6 used in conjunction provide guidelines to determine if a particular control device could and should be used for a given emission stream. Further design criteria (Chapter 4) and cost (Chapter 5) must be considered to enable a complete technical evaluation of the applicability of these devices to an emission stream.

3.3.1.1 Fabric Filters

Fabric filters, or baghouses, are an efficient means of separating particulate matter entrained in a gaseous stream. A fabric filter is typically least efficient collecting particles in the range of 0.1 to 0.3 μm diameter. For particles larger and smaller than this, a fabric filter can collect more than 99 percent of particles in an emissions stream. Fabric filters used to control emissions containing HAP's have two special constraints. First, they should have a closed, negative-pressure (suction) configuration to prevent accidental release of the gas stream and captured HAP's; and second, of the three principal fabric cleaning methods (i.e., mechanical shaking, reverse air flow, and pulse-jet cleaning), pulse-jet type cleaning is not recommended for HAP control situations.

As explained in Section 4.9, pulse-jet cleaned filters are not as efficient as those cleaned by other methods, and emissions are not as constant over a filtration cycle as those from filters using the other two cleaning methods. However, pulse-jet cleaning is widely used in general industrial fabric filter applications and, therefore, Section 4.9 does include information on pulse-jet cleaning. Further, Chapter 5 includes pulse-jet cost information to allow for review of permit applications indicating the use of pulse-jet baghouses.

Fabric filters using mechanical shaking or reverse air cleaning are fundamentally different from ESP's and venturi scrubbers in that they are not "efficiency" devices. A properly designed and operated fabric filter using one of these two cleaning methods will yield a relatively constant outlet particle concentration, regardless of inlet load changes. The typical outlet particle concentration range is between 0.003 to 0.01 grains/scf (gr/scf), averaging

Figure 3-6. Effluent characteristics for emission stream #5.

HAP EMISSION STREAM DATA FORM*

Company	Glaze Chemical Company	Plant Contact	Mr. John Leake
Location (Street)	87 Octane Drive	Telephone No.	(999) 555-5024
(City)	Somewhere	Agency Contact	Mr. Efrem Johnson
(State, Zip)		No. of Emission Streams Under Review	6

A. Emission Stream Number/Plant Identification #5 / Urea Evaporator Off-gas Exhaust

		(a)	(b)	(c)
B.	HAP Emission Source	evaporator off-gas	-	-
C.	Source Classification	process point	-	-
D.	Emission Stream HAP's	ammonia	-	-
E.	HAP Class and Form	inorganic vapor	-	-
F.	HAP Content (1,2,3)**	20,000 ppmv	-	-
G.	HAP Vapor Pressure (1,2)	8.46 atm. @ 68°F	-	-
H.	HAP Solubility (1,2)	provided	-	-
I.	HAP Adsorptive Prop. (1,2)	not given	-	-
J.	HAP Molecular Weight (1,2)	17 lb/lb-mole	-	-

K. Moisture Content (1,2,3)	2% vol.	P. Organic Content (1)***	
L. Temperature (1,2,3)	85°F	Q. Heat/O_2 Content (1)	/
M. Flow Rate (1,2,3)	3,000 scfm (max)	R. Particulate Content (3)	
N. Pressure (1,2)	atmospheric	S. Particle Mean Diam. (3)	
O. Halogen/Metals (1,2)	none / none	T. Drift Velocity/SO_3 (3)	/

U. Applicable Regulation(s)

V. Required Control Level: assume 98% removal

W. Selected Control Methods: absorption

*The data presented are for an emission stream (single or combined streams) prior to entry into the selected control method(s). Use extra forms if additional space is necessary (e.g., more than three HAP's). and note this need.

**The numbers in parentheses denote what data should be supplied depending on the data in steps "C" and "E":
1 = organic vapor process emission
2 = inorganic vapor process emission
3 = particulate process emission

***Organic emission stream combustibles less HAP combustibles shown on Lines D and F.

around 0.005 grains/scf. These numbers can be used to ascertain an expected performance level (see the following Example Case). This is not meant to be an absolute, definitive performance level. A vendor should assist in any attempt to quantify an actual performance level.

Variables important to achieve a given performance (i.e., air-to-cloth ratio, cleaning mechanism, fabric type) are discussed in detail in Section 4.9. Fabric filters are sensitive to emission stream temperature and a precooler or preheater may be required, as discussed previously. Fabric filters operate at low pressure drops, giving them low operating costs. They are generally not a feasible choice to control emission streams with a high moisture content.

3.3.1.2 Electrostatic Precipitators

Electrostatic precipitator particle removal occurs by charging the particles, collecting the particles, and transporting the collected particles into a hopper. ESP's are less sensitive to particle size than the other two devices, but very sensitive to those factors that affect the maximum electrical power (voltage) at which they operate. These are principally the aerosol density (grains/scf) and the electrical resistivity of the material. The electrical resistivity of the particles influences the drift velocity, or the attraction between the particles and the collecting plate. A high resistivity will cause a low drift velocity which will decrease the overall collection efficiency. Electrostatic precipitators are discussed further in Section 4.10.

3.3.1.3 Venturi Scrubbers

Venturi scrubbers use an aqueous stream to remove particulate matter from an emissions stream. The performance of a venturi scrubber is not affected by sticky, flammable, or corrosive particles. Venturi scrubbers are more sensitive to particle size distribution than either ESP or fabric filters. In general, venturi scrubbers perform most efficiently for particles above 0.5 μm in diameter (see Section 4.11 for further detail). Venturi scrubbers have a lower initial cost than either fabric filters or ESP's, but the high pressure drop required for high collection efficiencies contributes to high operating costs.

Example Case

Assume a facility is required to achieve an emission limit for particulate emissions from a municipal waste incinerator. The emissions stream particles consist primarily of fly ash; however, 10 percent of these particles is a HAP: cadmium. The characteristics of the emission stream after exiting a heat exchanger are shown in Figure 3-8. From Figure B.1-1 (Appendix B), the dew point of an emission stream containing 200 ppmv SO_3 and 5 percent moisture is approximately 327°F. Thus, the emission stream temperature (400°F) is within 50 to 100°F above its dew point, thus minimizing the amount of the HAP in vapor form and eliminating condensation problems.

Calculate the allowable outlet particle concentration:

$$\frac{110{,}000 \text{ acf}}{\text{min}} \times \frac{3.2 \text{ gr}}{\text{acf}} \times \frac{\text{lb}}{7{,}000 \text{ gr}} \times \frac{60 \text{ min}}{\text{hr}}$$

$$= 3{,}017 \frac{\text{lb}}{\text{hr}}$$

$$3{,}017 \frac{\text{lb}}{\text{hr}} \times (1 - 0.999) = 3.017 \frac{\text{lb}}{\text{hr}}$$

Since the HAP constitutes 10 percent of the total particulate matter, the outlet concentration of the HAP is:

$$0.10 \times 3.017 \frac{\text{lb}}{\text{hr}} = 0.3017 \frac{\text{lb HAP}}{\text{hr}}$$

This value assumes that the HAP is in particulate form and that it is collected as efficiently as the other particles.

A fabric filter will generally control particles to a limit of 0.005 gr/scf. Converting gr/scf to lb/hr yields:

$$\frac{0.005 \text{ gr}}{\text{scf}} \times \frac{110{,}000 \text{ acf}}{\text{min}} \times \frac{530°\text{R-scf}}{(400 + 460)°\text{R-acf}}$$

$$\times \frac{60 \text{ min-lb}}{7{,}000 \text{ gr-hr}} = 2.9 \frac{\text{lb}}{\text{hr}}$$

To calculate the HAP outlet emission rate:

$$2.9 \frac{\text{lb}}{\text{hr}} \times 0.10 = 0.29 \frac{\text{lb HAP}}{\text{hr}}$$

This value again assumes that the HAP is in particulate form and that it is collected as efficiently as the other gas stream particles. The calculated HAP emission rate from a fabric filter is less than the allowable rate, indicating that use of a fabric filter is an appropriate control technique for this emission stream.

In general, an electrostatic precipitator can achieve control efficiencies of 99.9 percent, provided the particle resistivity is not "high." The drift velocity of the particles (0.30 ft/sec) is indicative of particles with an "average" resistivity; therefore, an ESP can probably be used to control this stream and, thus, it also is an appropriate control technique for this emission stream.

A venturi scrubber has difficulty controlling particles below 0.5 μm diameter. The emission stream presented contains particles generally above this limit; therefore, a venturi scrubber may be an acceptable control device for this example and, thus, all three particulate control devices are appropriate control techniques for this emission stream. To determine the basic design parameters and actual applicability of each control device, Section 4.9 (Fabric Filter), Section 4.10 (ESP's), and Section 4.11 (Venturi Scrubbers) must be examined.

3.3.2 Control Techniques for Particulate Emissions from Fugitive Sources

Fugitive emission sources may be broken down into two source categories: process sources and area sources, as defined in Section 2.2. The methods used to control process sources of fugitive particulate emissions are generally different from those applied to area sources. Basically, process fugitive sources can employ conventional measures (i.e., capture techniques and add-on control devices) while area fugitive sources either cannot use conventional measures or the use of conventional measures is precluded due to cost. For example, fugitive emissions from unpaved roads cannot use conventional control measures, but fugitive emissions from area sources such as pumps and valves can be captured and ducted to a control device, although the costs may be prohibitive. Area sources are often controlled by preventive techniques rather than capture/control techniques.

Section 3.2.3 discusses methods of hooding and capture of process emissions. The remaining fugitive particulate emission control methodologies (i.e., nonconventional techniques) can be applied to multiple fugitive emission sources—both for process and area sources.

The following subsections discuss the different types of fugitive particulate emission control techniques that can be applied to general process and area fugitive particulate emission sources (i.e., sources common to many industries). (See Appendix A.8, reference 3, for industry-specific and source-specific information for fugitive particulate emission control and Appendix A.9, reference 3, for information on chemical dust suppressants.)

Figure 3-7. Effluent characteristics for emission stream #6.

HAP EMISSION STREAM DATA FORM*

Company	Glaze Chemical Company	Plant Contact	Mr. John Leake
Location (Street)	87 Octane Drive	Telephone No.	(999) 555-5024
(City)	Somewhere	Agency Contact	Mr. Efrem Johnson
(State, Zip)		No. of Emission Streams Under Review	6

A. Emission Stream Number/Plant Identification #6 / Styrene Recovery Condenser Vent

	(a)	(b)	(c)
B. HAP Emission Source	condenser vent	-	-
C. Source Classification	process point	-	-
D. Emission Stream HAP's	styrene	-	-
E. HAP Class and Form	organic vapor	-	-
F. HAP Content (1,2,3)**	13,000 ppmv	-	-
G. HAP Vapor Pressure (1,2)	provided	-	-
H. HAP Solubility (1,2)	insoluble in water	-	-
I. HAP Adsorptive Prop. (1,2)	not given	-	-
J. HAP Molecular Weight (1,2)	104 lb/lb-mole	-	-

K. Moisture Content (1,2,3)	negligible	P. Organic Content (1)***	none
L. Temperature (1,2,3)	90°F	Q. Heat/O_2 Content (1)	61.5 Btu/scf / 20.7% vol.
M. Flow Rate (1,2,3)	2,000 scfm (max)	R. Particulate Content (3)	
N. Pressure (1,2)	atmospheric	S. Particle Mean Diam. (3)	
O. Halogen/Metals (1,2)	none / none	T. Drift Velocity/SO_3 (3)	/

U. Applicable Regulation(s)

V. Required Control Level: assume 90% removal

W. Selected Control Methods: absorption, condensation

*The data presented are for an emission stream (single or combined streams) prior to entry into the selected control method(s). Use extra forms if additional space is necessary (e.g., more than three HAP's). and note this need.

**The numbers in parentheses denote what data should be supplied depending on the data in steps "C" and "E":
1 = organic vapor process emission
2 = inorganic vapor process emission
3 = particulate process emission

***Organic emission stream combustibles less HAP combustibles shown on Lines D and F.

An extensive review of available literature on fugitive emissions revealed that one reference included almost all necessary information pertinent to the scope of this handbook. Consequently, most of the following subsections are taken directly from the following document: *Technical Guidance for Control of Industrial Process Fugitive Particulate Emissions.*(16) An April 1985 draft final document from EPA on fugitive emissions (17) has been distributed and was reviewed. Although no significant changes were made to this section as a result of that document, it is referenced and does provide the reader with another comprehensive view of the subject.

Throughout the discussions, control efficiencies are stated for many of the control techniques. It is important to note that the efficiency values are estimates. The ability to quantify accurately the emission rates from a fugitive emission source has not yet been fully realized.

3.3.2.1 Process Fugitive Particulate Emission Control

Control of HAP process fugitive emissions may be accomplished by capturing the particulate material and venting it to an add-on control device (i.e., venturi scrubbers, fabric filters, and ESP's). Venting emissions is accomplished by exhausting particle-laden air through fixed or movable ducting under negative pressure. The airflow into the ducting must be sufficient to maintain particle size capture velocity and to overcome opposing air currents. The effect of opposing air currents can be eliminated by complete enclosure, or can be reduced by minimizing the opening of capture enclosures, or by utilizing curtains or partitions to block room air currents. If enclosure or installation of fixed hoods is not feasible due to space limitations or operational procedures, movable hoods may be a viable alternative. Movable hoods can be placed over the fugitive emissions source as the production cycle permits. For example, movable hoods can be

Table 3-6. Advantages and Disadvantages of Particulate Control Devices

	Advantages	Disadvantages
Baghouse	—Very efficient at removing fine particulate matter from a gaseous stream; control efficiency can exceed 99% for most applications. —Lower pressure drop than venturi scrubber when controlling fine particulates; i.e., 2" to 6" H_2O compared with ≥40" H_2O. —Can collect electrically resistive particles. —With mechanical shaking or reverse air cleaning, control efficiency is generally independent of inlet loading. —Simple to operate.	—Cannot control high temp stream (>550°F) without a precooler. —Cannot effectively control stream with high moisture content. —Highly erosive particles can damage the filter. —Mechanical collectors generally required upstream if significant amounts of large particulates (>20 μm) are present. —Needs special or selected fabrics to control corrosive streams. —Least efficient with particles between 0.1 μm to 0.3 μm diameter.
ESP	—Can control very small (<0.1 μm) particles with high efficiency. —Low operating costs with very low pressure drop (0.5" H_2O). —Can collect corrosive or tar mists. —Power requirements for continuous operation are low. —Wet ESP's can collect gaseous pollutants.	—High initial capital investment. —Not readily adaptable to changing conditions. —Conditioning agents may be necessary to control resistive particles. —More sensitive to particle loading than other two devices —Space requirements may be greater than that for a fabric filter or venturi scrubber.
Venturi	—Low initial investment. —Takes up relatively little space. —Can control sticky, flammable, or corrosive matter with few problems. —Can simultaneously collect particulates and gaseous matter. —Control efficiency is independent of particle resistivity. —Simple to operate, few moving parts.	—High operating cost due to high pressure drop. (40" H_2O or greater), particularly for smaller (<1 μm) particles. —Has wastewater and cleaning/disposal costs. —Least efficient with particles less than 0.5 μm diameter.

placed over the filling hatch in some types of trucks and rail cars during loading. Movable hoods have also been applied in the production of coke, whereby a movable hood follows quench cars during coke pushing. Another alternative is to evacuate an enclosed building to a control device.

Once the emissions are captured, the selection of the control device will be dependent on the emission stream characteristics, HAP characteristics, and the required performance levels (i.e., removal efficiency). It is important to remember that the required performance level for the control device is influenced by the efficiency of the capture system. The factors that affect the control device selection process are the same as for point sources, and are discussed in Section 3.3.1.

3.3.2.2 Area Fugitive Emission Control From Transfer and Conveying

Loss of material from conveyors is primarily at the feeding, transfer, and discharge points and occurs due to spillage or windage. The majority of particulate emissions are generally from spillage and mechanical agitation of the material at uncovered transfer points. However, emissions from inadequately enclosed systems can be quite extensive. Table 3-7 presents control techniques applicable to these emission sources.

Control by wet suppression methods includes the application of water, chemicals, and foam. The point of application is most commonly at the conveyor feed and discharge points, with some applications at conveyor transfer points. Wet suppression with water only is a relatively inexpensive technique; however, it has the inherent disadvantage of being short-lived. Control with chemicals (added to water for improved wetting) or foam is longer lasting but more expensive than water alone.

Foam is effective in dust suppression because small particles (in the range of 1 to 50 μm diameter) break the surface of the bubbles in the foam when they come in contact, thereby wetting the particles. Particles larger than 50 μm only move the bubbles away. The small wetted particles then must be brought together or brought in contact with larger particles to achieve agglomeration. If foam is injected into free-falling aggregate at a transfer point, the mechanical motion provides the required particle to bubble contact and subsequent particle-to-particle contact.

Highly diluted chemical wetting agents are applied by water jets ahead of any points in the conveying system where dusting occurs. The wetting agent breaks down the surface tension of the water, allowing it to spread further, penetrate deeper, and wet the small particles better than untreated water. With mechanical agitation of the material, the small particles agglomerate. For effective control, the

Figure 3-8. Effluent characteristics for a municipal incinerator emission stream.

HAP EMISSION STREAM DATA FORM*

Company	Incineration, Inc.	Plant Contact	Mr. Phil Brothers
Location (Street)	123 Main Street	Telephone No.	(999) 555-5624
(City)	Somewhere	Agency Contact	Mr. Ben Hold
(State, Zip)		No. of Emission Streams Under Review	1

A. Emission Stream Number/Plant Identification #1 / Incineration Exhaust

	(a)	(b)	(c)
B. HAP Emission Source	municipal incinerator	-	-
C. Source Classification	process point	-	-
D. Emission Stream HAP's	cadmium	-	-
E. HAP Class and Form	inorganic particulate	-	-
F. HAP Content (1,2,3)**	10%	-	-
G. HAP Vapor Pressure (1,2)		-	-
H. HAP Solubility (1,2)		-	-
I. HAP Adsorptive Prop. (1,2)		-	-
J. HAP Molecular Weight (1,2)		-	-

K. Moisture Content (1,2,3)	5% vol.	P. Organic Content (1)***	
L. Temperature (1,2,3)	400°F	Q. Heat/O_2 Content (1)	
M. Flow Rate (1,2,3)	110,000 scfm	R. Particulate Content (3)	3.2 grains/acf, flyash
N. Pressure (1,2)	atmospheric	S. Particle Mean Diam. (3)	1.0 μm
O. Halogen/Metals (1,2)	none / none	T. Drift Velocity/SO_3 (3)	0.30 ft/sec / 200 ppmv

U. Applicable Regulation(s)

V. Required Control Level assume 99.9% removal

W. Selected Control Methods fabric filter, electrostatic precipitator, venturi scrubber

*The data presented are for an emission stream (single or combined streams) prior to entry into the selected control method(s). Use extra forms if additional space is necessary (e.g., more than three HAP's). and note this need.

**The numbers in parentheses denote what data should be supplied depending on the data in steps "C" and "E":
1 = organic vapor process emission
2 = inorganic vapor process emission
3 = particulate process emission

***Organic emission stream combustibles less HAP combustibles shown on Lines D and F.

spray should be applied at each point where the particles might be fractured, allowed to free fall, or subjected to strong air currents.

3.3.2.3 Area Fugitive Emission Control From Loading and Unloading

Loading and unloading bulk material is common to many processing industries. Loading and unloading operations can be either for external transportation of material to or from a facility or for internal transportation within a facility (for example, internal transportation might consist of loading of a mining haul truck with ore via a front-end loader for subsequent unloading to a crushing process). (See reference 16 or Appendix A.8, reference 3, for industry-specific information on loading and unloading for internal transportation.)

Various control technology applications for loading and unloading operations are presented in Table 3-8. These techniques can be used alone or at times in various combinations. Generally, the simultaneous use of more than one technique will provide increased levels of control.

Rail car and truck loading—To minimize particulate emissions from rail car and truck loading, the entire operation can be enclosed by the use of doors on the loading shed. This prevents a wind tunnel effect and allows dust emitted in the enclosure to settle to the ground within the enclosure. By venting the entire enclosure to a control device, dust leakage around the doors and any other openings can be prevented, thus ensuring near 100 percent control.

Exhausting the car or truck body to a dust removal device reduces emissions if the body is fairly well enclosed. In open type rail or truck bodies this technique is not too effective.

Choke-feed eliminates free fall of material into the car or truck. In this technique the mouth of the feed tube is immersed in the material being unloaded. This technique only works for fairly free-flowing dry material. A telescopic chute or spout also essentially eliminates the free-fall distance of the material being loaded. This type of system can be used on all types of material. Both the choke-feed and telescopic chute methods are only partially effective in eliminating emissions since the surface of

Table 3-7. Control Technology Applications for Transfer and Conveying Sources (16)

Emission Points	Control Procedure
Conveyor System (belt, bucket elevator, etc.)	Enclosure 1. Top covered (marginal control) 2. Sides and top covered (good control) 3. Completely enclosed (excellent control) Wet suppression (water, chemical, foam) at conveyor feed points Belt scrapers and wipers Mechanical belt turnovers Replacement with pneumatic system or screw conveyor
Transfer and Transition Points	Enclosure Hoods, covers, or canopies with exhaust to removal equipment (fabric filters, and wet-collectors) Wet suppression (water, chemical, foam)

the loaded material is constantly disturbed by new material. This surface is subject to wind and dust entrainment.

Movable hoods exhausted to a dust removal system can be placed over the filling hatch in some types of trucks and railcars during loading. By keeping other openings on the body closed, any dust generated in loading must be emitted through the single open hatch. A hood with sufficient airflow mounted around this opening could capture most of the dust generated.

Wet suppression techniques, when applied to loading operations, can reduce airborne dust to some extent. The loading process naturally breaks up surface coatings, but some small dust particles will adhere to larger pieces so as not to become entrained. Many materials cannot be readily wetted and this technique could not be used for these materials.

Barge and ship loading—Due to their larger size, barge and ship loading present unique problems for dust control. However, a number of control techniques have been developed and utilized, especially at some of the larger shipping terminals.

The use of tarpaulins or similar covers over hatches on ships and enclosed barges reduces airborne emissions by preventing their escape. Air, displaced by the material being loaded, causes the hold to become slightly pressurized during loading, and the hold must be vented at some point if the hatches are air-tight. Thus, a more effective control system incorporates an exhaust system for the hold. This exhaust system is connected to a dust control system such as fabric filter with the collected material being returned to the hold. Such a system can practically eliminate loading emissions if carefully maintained and properly operated. The use of a canopy hood and exhaust system over the loading boom is less effective than a totally enclosed system, but can still reduce emissions and is a viable alternative for open barges. Effective utilization of this technique requires some type of wind break to increase the hood capture efficiency.

Choke feed and telescopic chutes or spouts as previously described can also be used for loading both enclosed and open ships or barges. Wet suppression techniques may also help reduce airborne emissions if the product specifications do not prohibit use of this technique.

Rail car and truck unloading—Many of the unloading dust control techniques are identical to the loading techniques. When a rail car or truck is tilted and materials are dumped into an underground chamber through a grating, exhausting air from this chamber through a control device will effectively reduce emissions. By causing air to flow down through the grating, dust emissions are contained. The face velocity of air through the grating is a critical design parameter in this technique. Unloading cars with a screw conveyor causes less distribution of the material and thereby less dust. Problems of material handling and time requirements limit the application of this technique. Pneumatic unloading of very fine materials is an effective and widely used technique that practically eliminates dust emissions. With this system, careful maintenance of hose fittings and the fabric filter through which the conveying air exhausts is required.

Barge and ship unloading—Control of barge and ship unloading requires enclosure of the receiving point on the shore and possibly exhausting of that enclosure to a control device. A good enclosure with an exhaust system can provide essentially 100 percent capture. For open ships and barges which use buckets and conveyors, a partially enclosed bucket will reduce windblown dust. When observation of the bucket by the operator is required, a transparent heavy plastic sheet can be used as a cover. This system is only partially effective and must usually be supplemented with other controls, such as tighter fitting covers, wind breaks, or possibly wet suppression.

3.3.2.4 Area Fugitive Emission Control From Paved and Unpaved Roads

Dust on the surface of paved roads is deposited by such processes as mud track-out on vehicle tires, atmospheric fallout, spillage or leakage from trucks, pavement wear and decomposition, runoff or wind erosion from adjacent land areas, deposition of biological debris, wear from tires and brake linings, and wear of anti-skid compounds. This material is reentrained by contact with tires and by the air turbulence created by passing vehicles.

Table 3-8. Control Technology Applications for Loading and Unloading Operations (16)

Emission Points	Control Procedures: Loading Operations	Control Procedures: Unloading Operations
Railcar, Truck	—Drive-through enclosure with doors at both ends —Exhaust of entire enclosure to dust removal equipment —Movable hood over hatch opening —Exhaust of car hopper to dust removal equipment —Choke-feed or telescopic chute to confine and limit free-fall distance (gravity loading) —Wet suppression (water, chemicals)	—Drive-through enclosure with doors at both ends —Exhaust of enclosure to dust removal equipment —Exhaust air from below grating of receiving hopper to removal equipment —Choke-feed to receiving pit (hopper car and hopper truck) —Unloading with screw conveyor (box car) —Wet suppression (water, chemicals) —Use of pneumatic unloading system
Barge, Ship	—Use of tarpaulins or covers over the holds —Canopy and exhaust system over the loading boom, with attached tarps around the hatch —Exhaust of ship hold to dust removal equipment —Choke-feed or telescopic chute to confine and limit free-fall distance —For tanker types, use of gravity filler spouts with concentric outer exhaust duct to control equipment —Wet suppression (water, chemicals)	—Enclosure of top of clamshell bucket with transparent material and maintenance of closure seals and teeth on bottom of bucket —Enclosure of shoreside receiving hopper —Exhaust of enclosed shoreside receiving hopper to dust removal equipment

On unpaved roads, the road base itself serves as the main source of dust. As with paved roads, the dust becomes airborne by contact with vehicles' tires and by air turbulence from passing vehicles. Also, some of the fugitive dust from unpaved roads is attributed to wind erosion. On both paved and unpaved roads, traffic movement causes the continuing mechanical breakdown of large particles on the road surface, thus providing new material in the suspended particulate size range. Available procedures for reducing emissions from plant roads and their estimated efficiencies are presented in Table 3-9.

Paved streets and roads in a plant area can be cleaned on a frequent schedule to reduce the amount of particulate material on the surface that is available for reentrainment. Flushers and vacuum-type motorized street cleaners are both quite effective in removing surface material and thereby reducing emission rates from vehicles using the cleaned streets. Because raw material accumulates rapidly on the streets, the overall effectiveness of a street-cleaning measure is a function of the frequency of cleaning and the removal efficiency of the equipment.

For plants with small amounts of paved roads, industrial vacuum sweepers or contracted sweeping programs (such as many shopping centers use) would be more appropriate than the larger vacuum street cleaning equipment used on public streets. Mechanical broom sweepers have been shown to be ineffective from an air pollution control standpoint in that they redistribute material into the active traffic lanes of the streets and they remove almost none of the fine material (less than 43 μm) that is subject to reentrainment.

Many street sweepers depend upon the material being concentrated in the gutter in order to achieve good collection efficiency, and therefore cannot be used on streets without curbs and gutters. However, the smaller industrial sweepers are usually designed for use in warehouse and storage areas that are not curbed. A factor which might limit the applicability of street flushers in plants is that unpaved areas adjacent to the streets would be wet by the water spray and then become subject to mud trackout onto the streets by equipment and vehicles driving through these areas.

Good housekeeping practices include the rapid removal of spills on roadways and at conveyor transfer points. Preventive measures include covering of truck beds to prevent windage losses, cleaning of truck tires and undercarriages to reduce mud trackout onto paved roads, and minimizing the pick-up of mud by trucks.

Table 3-9. Control Technology Applications for Plant Roads (16)

Emission Points	Control Procedure	Efficiency
Paved Streets	Street cleaning	No estimate
	Housecleaning programs to reduce deposition of material on streets	No estimate
	Vacuum street sweeping (daily) (2)	25% (17)
	Speed reduction	Variable
Unpaved Roads	Paving	85%
	Chemical stabilization	50%
	Watering	50%
	Speed reduction	Variable
	Oiling and double chip surface	85%
Road Shoulders	Stabilization	80%

The paving of unpaved roadways is the most permanent of the various types of controls. However, the degree of effectiveness of this technique is highly dependent on prevention of excessive surface dust loading.

Watering of unpaved roads is effective only when carried out on a regular basis. The schedule depends on climate, type of surface material, vehicle use, and type of vehicles.

Oiling unpaved roads is more effective than watering and needs to be applied less often. However, special precautions must always be taken so as not to add to surface water runoff problems.

3.3.2.5 Area Fugitive Emission Control From Storage Piles

Most dust arises from stockpile areas as the material is dumped from the conveyor or chute onto the pile, and as bulldozers move the pile. During periods with high wind speeds (greater than about 6 m/sec [13 mph]) or low moisture, wind erosion of a nonweathered surface may also cause emissions. Applicable control techniques for open storage piles are presented in Table 3-10.

Enclosing materials in storage is generally the most effective means of reducing emissions from this source category because it allows the emissions to be captured. However, storage bins or silos may be very expensive. Storage buildings must be designed to withstand wind and snow loads and to meet requirements for interior working conditions. One alternative to enclosure of all material is to screen the material prior to storage, sending the oversize material to open storage and the fines to silos.

Table 3-10. Control Technology Applications for Open Storage Piles (16)

Emission Points	Control Procedure	Efficiency
Loading onto Piles	Enclosure	70-99%
	Chemical wetting agents or foam	80-90%
	Adjustable chutes	75%
Movement of Pile	Enclosure	95-99%
	Chemical wetting agents	90%
	Watering	50%
	Traveling booms to distribute material	No estimate
Wind Erosion	Enclosure	95-99%
	Wind screens	Very low
	Chemical wetting agents or foam	90%
	Screening of material prior to storage, with fines sent directly to processing or to a storage silo	No estimate
Loadout	Water spraying	50%
	Gravity feed onto conveyor	80%
	Stacker/reclaimer	25-50%

Wind screens, or partial enclosure of storage piles, can reduce wind erosion losses but do not permit capture of the remaining storage pile fugitive emissions. Earthen berms, vegetation, or existing structures can serve as wind screens.

Telescoping chutes, flexible chute extensions, and traveling booms are used to minimize the free fall of material onto the pile and resulting emissions. Similarly, emissions due to loadout can be reduced by reclaiming the material from the bottom of the pile with a mechanical plow or hopper system. The use of telescoping chutes and flexible chute extensions for piles with high material flow rates may require closer control of operations because of the possibility of jamming. Traveling or adjustable booms can handle high flow rates, but have greater operating costs.

Wetting agents or foams that are sprayed onto the material during processing or at transfer points retain their effectiveness in subsequent storage operations. Wetting agents retain surface moisture for extended periods, thereby preventing dusting. Spraying of the material prior to storage may not be possible in cases where product contamination could result (e.g., Portland cement clinker) or where the material is water soluble. However, such materials are generally not placed in open storage anyway. Steam has also been found to be an effective dust suppressant for some short-term storage operations.

3.3.2.6 Area Fugitive Emission Control From Waste Disposal Sites

Fugitive dust can occur anywhere dusty waste material is dumped for disposal. This includes overburden piles, mining spoils, tailings, fly ash, bottom ash, catch from air pollution control equipment, process overload discharges, building demolition wastes, contaminated product, etc. Like open storage, emissions come from dumping and from wind erosion across unprotected surfaces. Since waste piles are generally not disturbed after dumping, there are no emissions from an activity comparable to loading out of the storage pile. However, there may be emissions from transporting the waste material on-site (if it is dry when it is produced) or from a reclamation process such as landfill covering associated with the waste disposal operation. If the surface of the waste material does not include a compound that provides cementation upon weathering, or if the surface is not compacted, or if an area of very little rainfall, wind erosion of fines can occur with winds greater than about 21 km per hour (13 mph). Table 3-11 presents control techniques for waste disposal sites.

Table 3-11. Control Technology Applications for Waste Disposal Sites (16)

Emission Points	Control Procedure	Efficiency
Handling	Keep material wet	100%
	Cover or enclose hauling	No estimate
	Minimize free fall of material	No estimate
Dumping	Spray bar at dump area	50%
	Minimal free fall of material	No estimate
	Semi-enclose bin	No estimate
Wind Erosion	Cover with dirt or stable material	100%
	Chemically stabilize	80%
	Revegetate	25%-100%
	Rapidly reclaim newly filled areas	No estimate
Grading	Water	50%

3.4 References

1. *Control Technology for Toxic and Hazardous Air Pollutants.* McFarland, A. R., ed. Illinois Institute for Environmental Quality. Chicago, Illinois. 1975.

2. Committee on Industrial Ventilation. *Industrial Ventilation: A Manual of Recommended Practice.* 17th Edition. Lansing, Michigan. 1982.

3. U.S. EPA. *Evaluation of Control Techniques for Hazardous Air Pollutants - Appendices.* EPA-600/7-86-009b (NTIS PB86-167/038/AS.)October 1985.

4. U.S. EPA. *Flexible Vinyl Coating and Printing Operations — Background Information for Proposed Standards.* EPA-450/3-81-016a. January 1983.

5. U.S. EPA. *Publication Rotogravure Printing — Background Information for Proposed Standards.* EPA-450/3-80-031a. October 1980.

6. U.S. EPA. *Measurement of Process Capture Efficiency. Draft Report of Laboratory Testing.* EPA Contract No. 68-03-3038. January 1984.

7. U.S. EPA. *VOC Emissions from Petroleum Refinery Wastewater Systems — Background Information for Proposed Standards.* Research Triangle Park, North Carolina. July 1984.

8. U.S. EPA. *Assessment of Atmospheric Emissions from Petroleum Refining — Appendix B: Detailed Results.* EPA-600/2-80-075c. April 1980.

9. U.S. EPA. *VOC Fugitive Emissions in Synthetic Organic Chemicals Manufacturing Industry — Background Information for Promulgated Standards.* EPA-450/3-80-033b. June 1982.

10. U.S. EPA. *Fugitive Emission Sources of Organic Compounds — Additional Information on Emissions, Emission Reductions, and Costs.* EPA-450/3-82-010. April 1982.

11. Wilkins, G. E., J. H. E. Stelling, and S. A. Shareef. *Monitoring and Maintenance Programs for Pumps and Valves in Petroleum and Chemical Processing Plants: Costs and Effects on Fugitive Emissions.* Presented at 6th World Congress on Air Quality, IUAPPA, Paris. May 1983.

12. Wilkins, G. E., and J. H. E. Stelling. *Monitoring and Maintenance Programs for Control of Fugitive Emissions from Pumps and Valves in Petroleum and Chemical Processing Plants.* Presented at 1984 Industrial Pollution Control Symposium, ASME. New Orleans, Louisiana. February 1984.

13. U.S. EPA. *VOC Fugitive Emission Predictive Model — User's Guide.* EPA-600/8-83-029. October 1983.

14. U.S. EPA. *Evaluation of the Efficiency of Industrial Flares. Background — Experimental Design — Facility.* EPA-600/2-83-070. August 1983.

15. *Environmental Engineers' Handbook, Volume II: Air Pollution.* Liptak, B. G., ed. Chilton Book Company. Radnor, Pennsylvania. 1974.

16. U.S. EPA. *Technical Guidance for Control of Industrial Process Fugitive Particulate Emissions.* EPA-450/3-77-010. March 1977.

17. U.S. EPA. *Identification, Assessment, and Control of Fugitive Particulate Emissions.* Final report. EPA Contract No. 68-02-3922. May 1986. Contact Mr. Dale Harmon of EPA at (919) 541-2429.

Chapter 4
HAP Control Techniques

4.1 Background

This section describes and illustrates the procedures used to calculate the basic design and operating variables of HAP control techniques in terms of commonly employed design principles and values. For each technique, the handbook provides: (a) a brief description of how the technique works, (b) definitions of input data required, and (c) a step-by-step calculation procedure showing where each number used in the procedure comes from and how it is to be used. The procedures described in this handbook will result in conservatively designed control systems. In instances in which less conservatively designed control systems might achieve the target control level, more detailed calculation procedures requiring compound-specific data would be needed. This level of specificity is beyond the scope of this handbook.

The data for the HAP emission stream to be controlled are taken from the HAP Emission Stream Data Form given in Chapter 2. In case of a permit evaluation, however, these data should be supplied by the applicant. The reviewer may wish to confirm the completeness of the applicant's data by referring to Chapters 2 and 3.

The step-by-step calculation procedures are illustrated for each control technique using data based on Emission Streams 1 through 7 described in Chapter 3. In permit reviews, the calculated values are compared to the values in the permit application to determine the adequacy of the applicant's proposed design. Appendices C.2 through C.11 contain blank calculation sheets to use in applying the calculations described for each control technique. If control systems costs are required, see Chapter 5.

4.2 Thermal Incineration

Thermal incineration (Figure 4-1) is a widely used air pollution control technique whereby organic vapors are oxidized at high temperatures. The most important variables to consider in thermal incinerator design are the combustion temperature and residence time because these design variables determine the incinerator's destruction efficiency. Further, at a given combustion temperature and residence time, destruction efficiency is also affected by the degree of turbulence, or mixing of the emission stream and hot combustion gases, in the incinerator. In addition, halogenated organics are more difficult to oxidize then unsubstituted organics; hence, the presence of halogenated compounds in the emission stream requires higher temperature and longer residence times for complete oxidation. Thermal incinerators can achieve a wide range of destruction efficiencies. This discussion focuses on efficiencies of 98 to 99+ percent.

The incinerator flue gases are discharged at high temperatures and contain valuable heat energy. Therefore, there is a strong economic incentive for heat recovery. Typical recovery methods include heat exchange between the flue gases and the emission stream and/or combustion air and use of the available heat for process heat requirements (e.g., recycling flue gases to the process, producing hot water or steam, etc.). In most thermal incinerator applications, the available enthalpy in the flue gases is used for preheating the emission stream. This discussion will be based on a thermal incineration system where the emission stream is preheated.

The incineration of emission streams containing organic vapors with halogen or sulfur components may create additional control requirements. For example, if sulfur and/or chlorine are present in the emission stream, the resulting flue gas will contain sulfur dioxide (SO_2) and/or hydrogen chloride (HCl). Depending upon the concentrations of these compounds in the flue gas and the applicable regulations, scrubbing may be required to reduce the concentrations of these compounds. The selection and design of scrubbing systems are discussed in Section 4.7.

In this subsection, the calculation procedure will be illustrated using Emission Stream 1 described in Chapter 3. Appendix C.3 provides worksheets for calculations.

Figure 4-1. Schematic diagram of a thermal incinerator system.

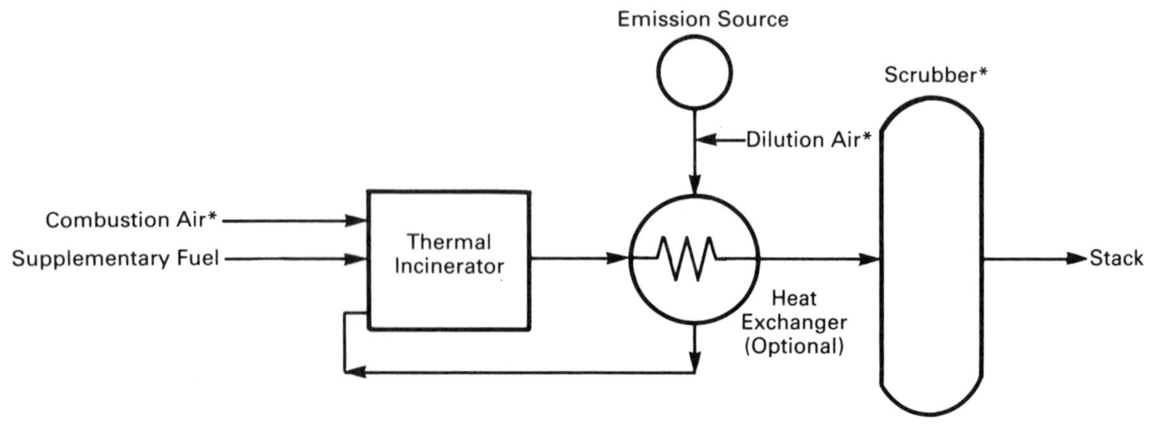

4.2.1 Data Required

The data necessary to perform the calculations consist of HAP emission stream characteristics previously compiled on the HAP Emission Stream Data Form and the required HAP control as determined by the applicable regulations.

Example Case

Maximum flow rate, Q_e = 15,000 scfm
Temperature, T_e = 120°F
Heat content, h_e = 0.4 Btu/scf
Oxygen content, O_2 = 20.6%
Moisture content, M_e = 2%
Halogenated organics: Yes ______ No __X__

Based on the control requirements for the emission stream:
Required destruction efficiency, DE = 99%

If dilution air is added to the emission stream upon exit from process, the data that will be used in the calculations are the resulting characteristics after dilution.

In the case of permit review for a thermal incinerator, the data outlined below should be supplied by the applicant; the calculations in this section will then be used to check the applicant's values.

Thermal incinerator system variables at standard conditions (70°F, 1 atm):

Reported destruction efficiency, $DE_{reported}$, %
Temperature of the emission stream entering the incinerator, T_e, °F (if no heat recovery);
T_{he}, °F (if a heat exchanger is employed)
Combustion temperature, T_c °F
Residence time, t_r, sec
Maximum emission stream flow rate, Q_e, scfm
Excess air, ex, %
Fuel heating value (assume natural gas), h_f, Btu/scf
Supplementary heat requirement, H_f, Btu/min
Combustion chamber volume, V_c, ft^3
Flue gas flow rate, Q_{fg}, scfm
Heat exchanger surface area (if a heat exchanger is employed), A, ft^2

4.2.2 Pretreatment of the Emission Stream: Dilution Air Requirements

In HAP emission streams containing oxygen/air and flammable vapors, the concentration of flammable vapors is generally limited to less than 25 percent of the lower explosive limit (LEL) to satisfy safety requirements imposed by insurance companies. (Note: The LEL for a flammable vapor is defined as the minimum concentration in air or oxygen at and above which the vapor burns upon contact with an ignition source and the flame spreads through the flammable gas mixture). In some cases, flammable vapor concentrations up to 40-50 percent of the LEL are permitted if on-line monitoring of VOC concentrations and automatic process control and shutdown are provided.

In general, emission streams treated by thermal incineration are dilute mixtures of VOC and air, and typically do not require dilution. For emission streams with oxygen concentrations greater than 16 percent and heat contents greater than 13 Btu-scf (corresponding to flammable vapor concentrations of approximately 25 percent of LEL), the calculation procedure in this handbook assumes that dilution air is required. (See Appendix B.2 for calculation of dilution air requirements and Appendix C.2 for a calculation worksheet.)

> **Example Case**
> Since O_2 = 20.6% and h_e = 0.4 Btu/scf, no dilution air is required.

4.2.3 Thermal Incinerator System Design Variables

Table 4-1 presents suggested combustion temperature and residence time values for thermal incinerators to achieve a given destruction efficiency. Two sets of values are shown in the table, one set for nonhalogenated emission streams and another set for halogenated emission streams. The combustion temperature and residence time values listed are conservative and assume adequate mixing of gases in the incinerator. The criteria in this table are not the only conditions for achieving the specified destruction efficiencies. For a given destruction efficiency, it may be possible to incinerate HAP emission streams at lower temperatures with longer residence times.

Based on the required destruction efficiency (DE), select appropriate values for T_c and t_r from Table 4-1.

Table 4-1. Thermal Incinerator System Design Variables (1,2)

Required	Nonhalogenated Stream		Halogenated Stream	
Destruction Efficiency DE (%)	Combustion Temperature T_c (°F)	Residence Time t_r (sec)	Combustion Temperature T_c (°F)	Residence Time t_r (sec)
98	1,600	0.75	2,000	1.0
99	1,800	0.75	2,200	1.0

> **Example Case**
> The required destruction efficiency is 99% and the HAP emission stream is nonhalogenated, therefore:
>
> T_c = 1,800°F (Table 4-1)
> t_r = 0.75 sec (Table 4-1)

In a permit evaluation, if the reported values for T_c and t_r are sufficient to achieve the required DE (compare the applicant's values with the values from Table 4-1), proceed with the calculations. If the reported values for T_c and t_r are not sufficient, the applicant's design is unacceptable. The reviewer may then wish to use the values for T_c and t_r from Table 4-1. (Note: If DE is less than 98 percent, obtain information from literature and incinerator vendors to determine appropriate values for T_c and t_r.)

4.2.4 Determination of Incinerator Operating Variables

4.2.4.1 Supplementary Heat Requirements

Supplementary fuel is added to the thermal incinerator to attain the desired combustion temperature (T_c). For a given combustion temperature, the quantity of heat needed to maintain the combustion temperature in the thermal incinerator is provided by: (a) the heat supplied from the combustion of supplementary fuel, (b) the heat generated from the combustion of hydrocarbons in the emission stream, (c) the sensible heat contained in the emission stream as it leaves the emission source, and (d) the sensible heat gained by the emission stream through heat exchange with hot flue gases. Typically, items (b) and (c) are very small and can often be neglected.

In general, emission streams treated by thermal incineration are dilute mixtures of VOC and air, and typically do not require additional combustion air. For preliminary calculations, it can be assumed that no additional combustion air is required if the oxygen concentration of the emission stream exceeds 16 percent. Depending on the heat content of the emission stream and the desired combustion temperature, combustion air requirements may be zero even when the oxygen concentration is below 16 percent. Hence, this cut-off value will lead to a conservative design. Use the following simplified equation for dilute streams to calculate supplementary heat requirements (based on natural gas):

$$H_f = 1.1\, h_f \left[\frac{Q_e(1 + 0.002\, M_e)\,[Cp_{air}(T_c - T_r) - Cp_{air}(T_{he} - T_r) - h_e]}{h_f - 1.4\, Cp_{air}\,(T_c - T_r)}\right] \qquad (4.2\text{-}1)$$

where:

H_f = supplementary heat requirement, Btu/min
h_f = heating value of natural gas, Btu/scf
Q_e = maximum emission stream flow rate, scfm
M_e = moisture content of the emission stream, %
Cp_{air} = average specific heat of air over a given temperature interval (T_r to T), Btu/scf-°F
T_c = combustion temperature, °F
T_r = reference temperature, = 70°F
T_{he} = emission stream temperature after heat recovery, °F
h_e = heat content of the emission stream, Btu/scf

Calculate T_{he} using the following expression if the value for T_{he} is not specified:

$$T_{he} = (HR/100)T_c + [1 - (HR/100)]\, T_e$$

where:
HR = heat recovery in the exchanger, %

Assume a value of 50 percent if no other information is available.

The factor 1.1 in Equation 4.2-1 accounts for an estimated heat loss of 10 percent in the incinerator. Supplementary heat requirements are based on maximum emission stream flow rate, and hence will lead to a conservative design. In this handbook, it is assumed that the minimum supplementary heat requirement is 5 Btu/min per scfm of emission stream.

A graph of Equation 4.2-1 is shown in Figure 4-2, where the ratio H_f/Q_e is plotted against the emission stream heat content (h_e) for four different combustion temperatures (T_c). Instead of evaluating Equation 4.2-1, Figure 4-2 can be used directly to determine supplementary heat requirements. This graph is based on the following assumptions: (1) temperature of the emission stream (T_e) is 100°F, (2) moisture content of the emission stream (M_e) is 2 percent, (3) preheat temperature of the emission stream (T_{he}) is based on 50 percent heat recovery in the heat exchanger, and (4) h_f = 882 Btu/scf (based on the lower heating value for natural gas).

Figure 4-2. Supplementary heat requirement vs. emission stream heat content (dilute stream/no combustion air).

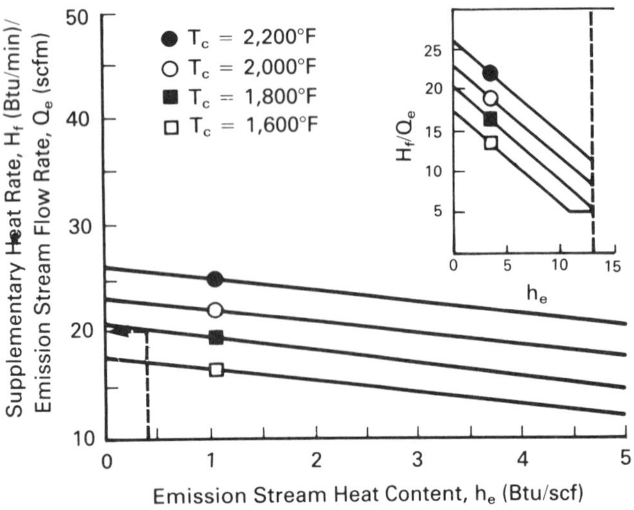

For emission streams that do not contain sufficient quantities of oxygen to satisfy the combustion air requirements (e.g., process emissions), refer to Figure 4-3 which shows a plot of H_f/Q_e versus h_e to obtain a conservative estimate for H_f. Figure 4-3 is based on the same assumptions as those stated for Figure 4-2. In addition, the oxygen content of the emission stream (O_2) is assumed as zero; this corresponds to maximum combustion air requirements for the thermal incinerator system. If the oxygen content of a particular emission stream falls between zero and 16 percent, use Figure 4-3 to obtain a conservative estimate of H_f/Q_e.

Figure 4-3. Supplementary heat requirement vs. emission stream heat content (no oxygen in emission stream/maximum combustion air).

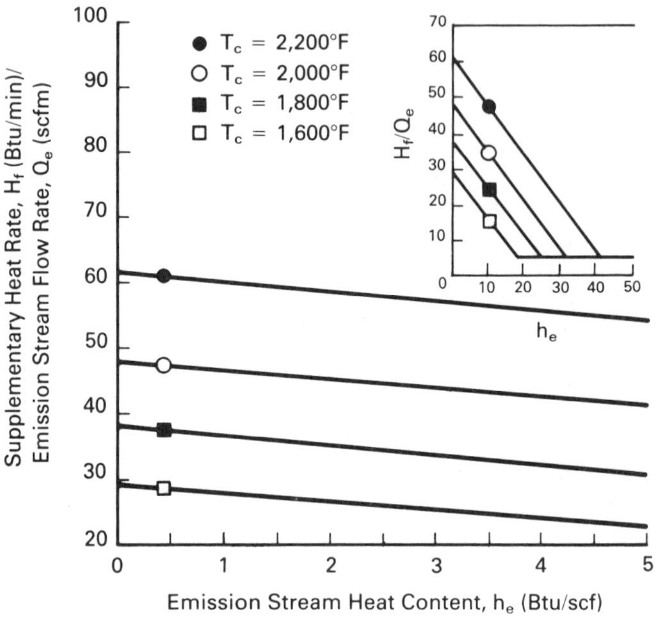

Example Case

Using Equation 4-1:

Since the emission stream is very dilute and has an oxygen content greater than 16%, Equation 4.2-1 is applicable. The values to be inserted in the equation are:

Q_e = 15,000 scfm
M_e = 2%
h_e = 0.4 Btu/scf
T_c = 1,800°F (Table 4-1)
T_r = 70°F
T_{he} = 960°F (based on heat recovery of 50%)
Cp_{air} = 0.0196 Btu/scf-°F for the interval 70° - 1,800°F (reference 3)
Cp_{air} = 0.0187 Btu/scf-°F for the interval 70° - 960°F (reference 3)
h_f = 882 Btu/scf

$$H_f = 1.1 \times 882 \left[\frac{[15{,}000\,(33.91 - 16.64 - 0.4)]}{882 - 47.5} \right]$$

H_f = 295,000 Btu/min

(Note: H_f is greater than the minimum supplementary heat requirement assumed in this handbook.)

Using Figure 4-2:

For h_e = 0.4 Btu/scf and T_c = 1,800°F, H_f/Q_e as indicated in the figure is about 20 Btu/min/scfm. Multiplying 20 by Q_e, (20 x 15,000), yields an approximate value of 300,000 Btu/min for H_f.

4.2.4.2 Flue Gas Flow Rate

Flue gas is generated as a result of the combustion of supplementary fuel and the emission stream. Use the following equation to calculate flue gas flow rate:

$$Q_{fg} = Q_e + Q_f + Q_c \tag{4.2-2}$$

where:
Q_{fg} = flue gas flow rate, scfm
Q_f = natural gas flow rate, scfm
Q_c = combustion air requirement, scfm

Calculate Q_f from the following equation:

$$Q_f = H_f/h_f \tag{4.2-3}$$

As indicated earlier, Q_c will typically be zero for dilute emission streams with oxygen contents (O_2) greater than 16 percent. If O_2 is less than 16 percent, then:

$$Q_c = (0.01\ H_e + 9.4\ Q_f)\ (1 + 0.01\ ex) - 0.0476\ O_2\ Q_e$$

where:
H_e = heat generated due to the combustion of hydrocarbons in the emission stream
$H_e = Q_e h_e$

Example Case

Using Equation 4.2-3:
H_f = 295,000 Btu/min
h_f = 882 Btu/scf
Q_f = 295,000/882 = 330 scfm

Using Equation 4.2-2:
Q_e = 15,000 scfm
Q_c = O (since O_2 is greater than 16%)
Q_{fg} = 15,000 + 330 + 0
Q_{fg} = 15,330 scfm

4.2.5 Combustion Chamber Volume

The flue gas flow rate (Q_{fg}) determined by Equation 4.2-2 is expressed at standard conditions. In order to calculate the combustion chamber size, Q_{fg} must be expressed at actual conditions, i.e., temperature effects must be considered (assume pressure effects are negligible). Use the following equation to convert Q_{fg} from "scfm" to "acfm":

$$Q_{fg,a} = Q_{fg}\ [(T_c + 460)/530] \tag{4.2-4}$$

where:
$Q_{fg,a}$ = flue gas flow rate at actual conditions, acfm

The volume of the combustion chamber (V_c) is determined from the residence time (t_r) and flue gas flow rate at actual conditions ($Q_{fg,a}$) according to the following equation:

$$V_c = [(Q_{fg,a}/60)\ t_r] \times 1.05 \tag{4.2-5}$$

The factor 1.05 is used in Equation 4.2-5 to increase the chamber volume by 5 percent. This technique is an accepted industry practice and allows for fluctuations in the operating conditions (e.g., flowrate, temperature, etc.). The smallest commercially available incinerator has a combustion chamber volume of about 36 ft^3 (1 m^3). If the calculated V_c is less than 36 ft^3, define V_c as 36 ft^3.

Example Case

Using Equation 4.2-4:
T_c = 1,800°F
Q_{fg} = 15,330 scfm
$Q_{fg,a}$ = 15,330 [(1,800 + 460)/530]
$Q_{fg,a}$ = 65,370 acfm

Using Equation 4.2-5:
t_r = 0.75 sec (from Table 4-1)
V_c = [(65,370/60) 0.75] 1.05
V_c = 860 ft^3

4.2.6 Heat Exchanger Size

The size of the heat exchanger required for preheating the emission stream to T_{he} before it enters the thermal incinerator is based on the heat exchanger design. Use the following expression to calculate the required size, i.e., surface area, of the heat exchanger:

$$A = \frac{[60\ Q_e\ (1 + 0.002\ M_e)\ Cp_{air}\ (T_{he} - T_e)]}{U \Delta T_{LM}} \tag{4.2-6}$$

where:
A = heat exchanger surface area, ft^2
T_e = emission stream temperature, °F
U = overall heat transfer coefficient, Btu/hr-ft^2-°F
ΔT_{LM} = logarithmic mean temperature difference, °F

and:

$$\Delta T_{LM} = [(T_c - T_{he}) - (T_{hc} - T_e)]\ \ln\ [(T_c - T_{he})/(T_{hc} - T_e)]$$

For dilute emission streams that do not require additional combustion air, ΔT_{LM} can be approximated by:

$$\Delta T_{LM} = T_c - T_{he}$$

For a recuperative heat exchanger where the heat transfer takes place between two gas streams, the overall heat transfer coefficient (U) ranges from 2 to 8 Btu/hr-ft°F, generally depending on the heat exchanger configuration and properties of the gas streams.

Equation 4.2-6 has been evaluated for dilute emission streams that require no additional combustion air, as shown in Figure 4-4. In the figure, heat exchanger surface area (A) is plotted against the emission stream flow rate (Q_e). The assumptions inherent in this figure are the same as those described for Figure 4-2. The overall heat transfer coefficient is assumed as 4 Btu/hr-ft^2-°F.

Example Case

Using Equation 4.2-6:

Q_e = 15,000 scfm
M_e = 2%
T_{he} = 960°F (based on heat recovery of 50%)
T_e = 120°F (input data)
Cp_{air} = 0.0187 Btu/scf-°F for the interval 120° - 960°F (reference 3)
U = 4 Btu/hr-ft^2-°F
ΔT_{LM} = T_c - T_{he}
T_c = 1,800°F
ΔT_{LM} = 1,800° - 960° = 840°F

Substituting in Equation 4.2-6:

$$A = \frac{[60 \times 15{,}000 \times 1.004 \times 0.0187\,(960 - 120)]}{(4 \times 840)}$$

$$A = 4{,}200 \text{ ft}^2$$

Using Figure 4-4:

For Q_e = 15,000, the value for A from the figure is about 4,000 ft^2.

For emission streams that are not dilute and require additional combustion air, use Figure 4-5 to obtain an estimate of the heat exchanger surface area. In Figure 4-5, the ratio A/Q_e is plotted against the emission stream content (h_e) for four different combustion temperature (T_c). The assumptions inherent in this figure are the same as those stated for Figure 4-4; in addition, maximum combustion air requirements are assumed (i.e., O_2 = 0). If the conditions represented in Figures 4-4 and 4-5 are not directly applicable for a particular emission stream, use Figure 4-4 to obtain a conservative estimate.

4.2.7 Evaluation of Permit Application

Using Table 4-2, compare the results from the calculations and the values supplied by the permit applicant. The calculated values in the table are based on the example. The value of the combustion chamber volume (V_c) is determined indirectly from the flue gas flow rate (Q_{fg}), and Q_{fg} is determined from the emission stream flow rate (Q_e), combustion air requirement (Q_c), and supplementary fuel requirement (Q_f). Therefore, if there are differences between the calculated and reported values for Q_{fg} and V_c, these will be dependent on the differences between the calculated and reported values for Q_c and Q_f. If the calculated values for Q_c, H_f, Q_f, and A differ from the reported values for these variables, the differences may be due to the assumptions involved in the calculations. Therefore, further discussions with the permit applicant will be necessary to find out about the details of the design and operation of the proposed thermal incinerator system.

If the calculated values and the reported values are not different, then the design and operation of the proposed thermal incinerator system may be considered appropriate based on the assumptions used in this handbook.

Figure 4-4. Heat exchanger size vs. emission stream flow rate (dilute stream/no combustion air).

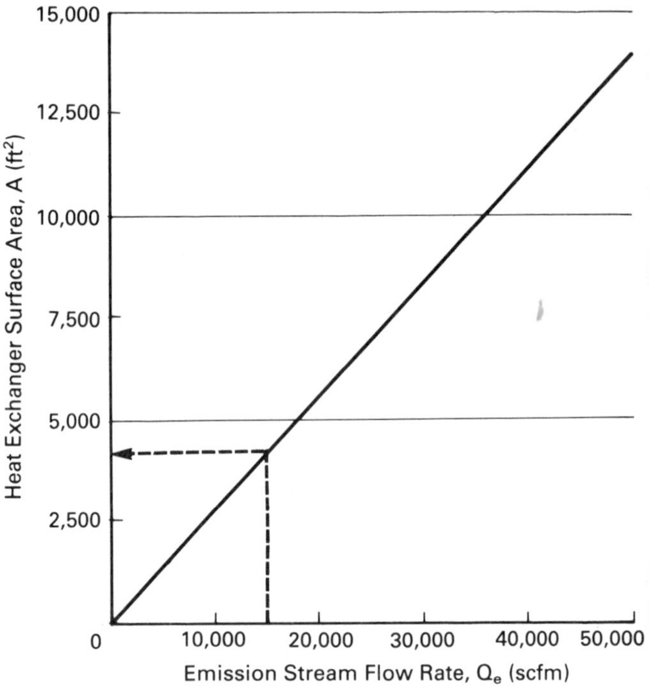

Figure 4-5. Heat exchanger size vs. emission stream heat content (no oxygen in emission stream/maximum combustion air).

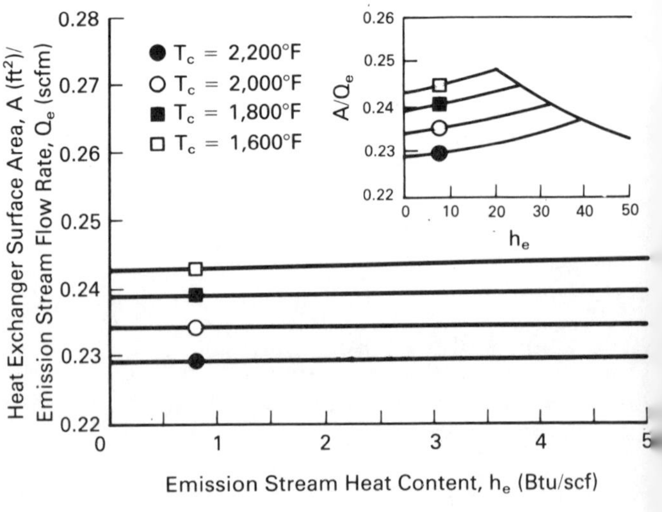

Table 4-2. Comparison of Calculated Values and Values Supplied by the Permit Applicant for Thermal Incineration

	Calculated Value (Example Case)[a]	Reported Value
Supplementary heat requirement, H_f	295,000	...
Supplementary fuel flow rate, Q_f	330 scfm	...
Flue gas flow rate, Q_{fg}	15,330 scfm	...
Combustion chamber size, V_c	860 ft^3	...
Heat exchanger surface area, A	4,200 ft^2	...

[a] Based on Emission Stream 1.

4.3 Catalytic Incineration

Catalytic incineration (Figure 4-6) is an air pollution control technique whereby VOC's in an emission stream are oxidized with the help of a catalyst. A catalyst is a substance that accelerates the rate of a reaction at a given temperature without being appreciably changed during the reaction. Catalysts typically used for VOC incineration include platinum and palladium; other formulations are also used, including metal oxides for emission streams containing chlorinated compounds. The catalyst bed (or matrix) in the incinerator is generally a metal mesh-mat, ceramic honeycomb, or other ceramic matrix structure designed to maximize catalyst surface area. The catalysts may also be in the form of spheres or pellets. Before passing through the catalyst bed, the emission stream is preheated, if necessary, in a natural gas-fired preheater.

The performance of a catalytic incinerator is affected by several factors including: (a) operating temperature, (b) space velocity (reciprocal of residence time), (c) VOC composition and concentration, (d) catalyst properties, and (e) presence of poisons/inhibitors in the emission stream. In catalytic incinerator design, the important variables are the operating temperature at the catalyst bed inlet and the space velocity. The operating temperature for a particular destruction efficiency is dependent on the concentration and composition of the VOC in the emission stream and the type of catalyst used.

Space velocity is defined as the volumetric flow rate of the combined gas stream (i.e., emission stream + supplemental fuel + combustion air) entering the catalyst bed divided by the volume of the catalyst bed. As such, space velocity also depends on the type of catalyst used. At a given space velocity, increasing the operating temperature at the inlet of the catalyst bed increases the destruction efficiency. At a given operating temperature, as space velocity is decreased (i.e., as residence time in the catalyst bed increases), destruction efficiency increases. Catalytic incinerators can achieve overall VOC destruction efficiencies up to about 98 percent and HAP destruction efficiencies up to about 95 percent with space velocities in the range 30,000 to 100,000 hr^{-1}.(4,5,6) However, the greater catalyst volumes and/or higher temperatures required for higher destruction efficiencies (i.e., 99 percent) may make catalytic incineration uneconomical. This discussion will be based on HAP destruction efficiencies of 90 and 95 percent.

The performance of catalytic incinerators is sensitive to pollutant characteristics and process conditions. In the following discussion, it is assumed that the emission stream is free from poisons/inhibitors such as phosphorus, lead, bismuth, arsenic, antimony, mercury, iron oxide, tin, zinc, sulfur, and halogens. (Note: Some catalysts can handle emission streams containing halogenated compounds.)

Figure 4-6. Schematic diagram of a catalytic incinerator system.

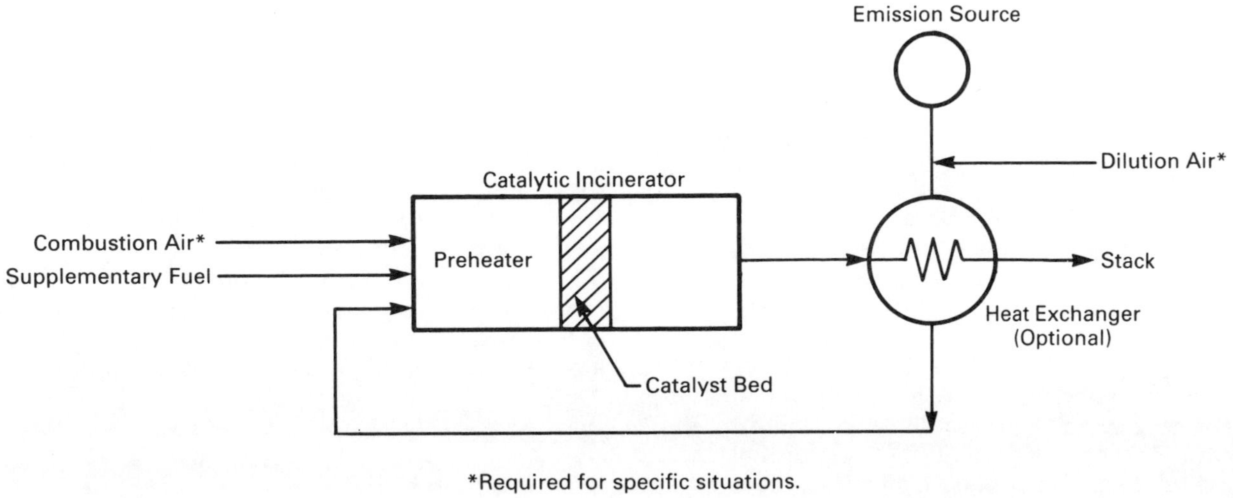

It is also assumed that the fluctuations in process conditions (e.g., changes in VOC content) are kept to a minimum.

The energy in the flue gases leaving the catalyst bed may be recovered in several ways including: (a) use of a recuperative heat exchanger to preheat the emission stream and/or combustion air, or (b) by use of the available energy for process heat requirements (e.g., recycling flue gases to the process, producing hot water or steam, etc.). In recuperative heat exchange, only a limited preheat is possible due to the temperature rise across the catalyst bed as a result of the combustion of VOC in the emission stream. High preheat temperatures accompanied by an increase in temperature due to combustion result in high operating temperatures at the catalyst bed, causing the catalyst bed to overheat and eventually lose its activity.

The following discussion will be based on fixed bed catalytic incinerator systems with no heat recovery and with recuperative heat exchange (i.e., preheating the emission stream). The calculation procedure will be illustrated using Emission Stream 2 described in Chapter 3. Appendix C.4 provides worksheets for calculations.

4.3.1 Data Required

The data necessary to perform the calculations consist of HAP emission stream characteristics previously compiled on the HAP Emission Stream Data Form and the required HAP control as determined by the applicable regulations.

> **Example Case**
> Maximum flow rate, Q_e = 20,000 scfm
> Temperature, T_e = 120°F
> Heat content, h_e = 2.1 Btu/scf
> Oxygen content, O_2 = 20.6%
> Moisture content, M_e = 2%
>
> Based on the control requirements for the emission stream:
> Required destruction efficiency, DE = 95%

If dilution air is added to the emission stream upon exit from the process, the data that will be used in the calculations are the resulting characteristics after dilution.

In the case of a permit review for a catalytic incinerator, the following data should be supplied by the applicant. The calculations in this section will then be used to check the applicant's values.

Catalytic incinerator system variables at standard conditions (70°F, 1 atm):

Reported destruction efficiency, $DE_{reported}$, %
Temperature of the emission stream entering the incinerator, T_e, °F (if no heat recovery); T_{he}, °F (if emission stream is preheated)
Temperature of flue gas leaving the catalyst bed, T_{co}, °F
Temperature of combined gas stream (emission stream + supplementary fuel combustion products) entering the catalyst bed,[a] T_{ci}, °F
Space velocity, SV, hr^{-1}
Supplementary heat requirement, H_f, Btu/min
Flow rate of combined gas stream entering the catalyst bed,[a] Q_{com}, scfm
Combustion air flow rate, Q_c, scfm
Excess air, ex, %
Catalyst bed requirement, V_{bed}, ft^3
Fuel heating value, h_f, Btu/scf
Heat exchanger surface area (if a heat exchanger is employed), A, ft^2

4.3.2 Pretreatment of the Emission Stream: Dilution Air Requirements

In general, catalytic incineration is applied to dilute emission streams. If emission streams with high VOC concentrations (i.e., heat content above 10 Btu/scf for air + VOC mixtures and above 15 Btu/scf for inert + VOC mixtures) are treated by catalytic incineration, they may generate enough heat upon combustion to deactivate the catalyst. Therefore, dilution of the emission stream with air is necessary to reduce the concentration of the VOC's.

Typically, the concentration of flammable vapors in HAP emission streams containing air is limited to less than 25 percent of the LEL (corresponding to a heat content of 13 Btu/scf) for safety requirements. In order to meet the safety requirements and to prevent damage to the catalyst bed, it is assumed in this handbook that catalytic incineration is applicable if the heat content of the emission stream (air + VOC) is less than or equal to 10 Btu/scf. For emission streams that are mixtures of inert gases and VOC (i.e., containing no oxygen), it is assumed that catalytic incineration is applicable if the heat content of the emission stream is less than or equal to 15 Btu/scf. Otherwise, dilution air will be required to reduce the heat content to levels below these cut-off values (i.e., 10 and 15 Btu/scf). For emission streams that cannot be characterized as air + VOC or inert + VOC mixtures, apply the 10

[a]If no supplementary fuel is used, the value for this variable will be the same as that for the emission stream.

Btu/scf cut-off value for determining dilution air requirements. See Appendix B.2 for calculating dilution air requirements.

> **Example Case**
> Since the heat content of the emission stream (h_e) is 2.1 Btu/scf, no dilution is necessary.

4.3.3 Catalytic Incinerator System Design Variables

Table 4-3 presents suggested values and limits for the design variables of a fixed bed catalytic incinerator system to achieve a given destruction efficiency. For specific applications, other temperatures and space velocities may be appropriate depending on the type of catalyst employed and the emission stream characteristics (i.e., composition and concentration). For example, the temperature of the flue gas leaving the catalyst bed may be lower than 1,000°F for emission streams containing easily oxidized compounds and still achieve the desired destruction efficiency. (See reference 7 or Appendix B.6, reference 8, for data on temperatures typically required for specific destruction efficiency levels for several compounds.)

Table 4-3. Catalytic Incinerator System Design Variables (1,5,6)

Required Destruction Efficiency DE(%)	Temperature at the Catalyst Bed Inlet[a] T_{ci} (°F)	Temperature at the Catalyst Bed Outlet[b] T_{co} (°F)	Space Velocity SV (hr^{-1})
90	600	1,000-1,200	40,000[c]
95	600	1,000-1,200	30,000[d]

[a] Minimum temperature of combined gas stream (emission stream + supplementary fuel combustion products) entering the catalyst bed is designated as 600°F to ensure an adequate initial reaction rate.

[b] Minimum temperature of flue gas leaving the catalyst bed is designated as 1,000°F to ensure an adequate overall reaction rate to achieve the required destruction efficiency. Note that this is a conservative value; it is in general a function of the HAP concentration (or heat content) and a temperature lower than 1,000°F may be sufficient to achieve the required destruction level. Maximum temperature of flue gas leaving the catalyst bed is limited to 1,200°F to prevent catalyst deactivation by overheating.

[c] Corresponds to 1.5 ft^3 of catalyst per 1,000 scfm of combined gas stream.

[d] Corresponds to 2.0 ft^3 of catalyst per 1,000 scfm of combined gas stream.

Note that the destruction efficiency for a given compound may vary depending on whether the compound is the only VOC in the emission stream, or it is part of a mixture of VOC's.(4) The destruction efficiency for a given compound in different VOC mixtures may also vary with mixture composition. (See reference 4 or Appendix B.6, reference 8, for compound-specific destruction efficiency data for two different VOC mixtures.)

Based on the required destruction efficiency (DE), specify the appropriate ranges for T_{ci}, T_{co}, and select the value for SV from Table 4-3.

> **Example Case**
> The required destruction efficiency is 95%; therefore:
> T_{ci} (minimum) = 600°F
> T_{co} (minimum) = 1,000°F
> T_{co} (maximum) = 1,200°F
> SV = 30,000 hr^{-1}

In a permit evaluation, determine if the reported values for T_{ci}, T_{co}, and SV are appropriate to achieve the required destruction efficiency by comparing the applicant's values with the values in Table 4-3. The reported value for T_{ci} should equal or exceed 600°F in order to obtain an adequate initial reaction rate. To ensure that an adequate overall reaction rate can be achieved to give the desired destruction efficiency without damaging the catalyst, check whether T_{co} falls in the interval 1,000° - 1,200°F. Note that 1,000°F is a conservative value. Then check if the reported value for SV is equal to or less than the value in Table 4-3. If the reported values are appropriate, proceed with the calculations. In some cases it may be possible to achieve the desired destruction efficiency at a lower temperature level. Otherwise, the applicant's design is considered unacceptable. In such a case, the reviewer may then wish to use the values in Table 4-3.

4.3.4 Determination of Incinerator System Variables

4.3.4.1 Supplementary Heat Requirements

Supplementary fuel is added to the catalytic incinerator system to provide the heat necessary to bring the emission stream up to the required catalytic oxidation temperature (T_{ci}) for the desired level of destruction efficiency. For a given T_{ci}, the quantity of supplementary heat needed is provided by: (a) the heat supplied from the combustion of supplementary fuel, (b) the sensible heat contained in the emission stream as it enters the catalytic incinerator system, and (c) the sensible heat gained by the emission stream through heat exchange with hot flue gases. If recuperative heat exchange is not practiced at a facility, then item (c) will be zero.

As mentioned earlier, emission streams treated by catalytic incineration are dilute mixtures of VOC and air, and typically do not require additional combustion air. As a conservative cut-off value, it can be assumed that no additional combustion air is required if the emission stream oxygen content (O_2) is greater than or equal to 16 percent.

Before calculating the supplementary heat requirements, the temperature of the flue gas leaving the catalyst bed (T_{co}) should be estimated to ensure that an adequate overall reaction rate can be achieved to give the desired destruction efficiency without damaging the catalyst. In other words, check whether T_{co} falls in the interval 1,000° - 1,200°F. Use the following expression to calculate T_{co}, taking into consideration the temperature rise across the catalyst bed due to heat generation from combustion of VOC in the emission stream:

$$T_{co} = T_{ci} + 50\ h_e \qquad (4.3\text{-}1)$$

where:

h_e = heat content of the emission stream, Btu/scf

In this expression, it is assumed that the heat content of the emission stream and the combined gas stream is the same. Inserting T_{ci} = 600°F, if T_{co} is in the range 1,000° - 1,200°F, then T_{ci} = 600°F is satisfactory. If T_{co} is less than 1,000°F, use the following equation to determine an appropriate value for T_{ci} (above 600°F) and use this value in the following calculations:

$$T_{ci} = 1{,}000 - 50\ h_e \qquad (4.3\text{-}2)$$

(Note: Emission streams with high heat contents will be diluted based on the requirements discussed in Section 4.3.2. Therefore, values for T_{co} exceeding 1,200°F should not occur.)

To calculate supplementary heat requirements (based on natural gas as the fuel), use the following simplified equation for dilute emission streams that require no additional combustion air:

$$H_f = 1.1\ h_f \times Q_e(1 + 0.002\ M_e) \left[\frac{Cp_{air}\,(T_{ci}-T_r) - Cp_{air}\,(T_{he} - T_r)}{h_f^{-1.4}\ Cp_{air}\,(T_{ci} - T_r)}\right] \qquad (4.3)$$

where:

H_f = supplementary heat requirement, Btu/min
h_f = heating value of natural gas, Btu/scf
Q_e = maximum emission stream flow rate, scfm
M_e = moisture content of the emission stream, percent
Cp_{air} = average specific heat of air over a given temperature interval, (T_r -T) Btu/scf-°F
T_{ci} = temperature of combined gas stream entering the catalyst bed, °F
T_r = reference temperature, = 70°F
T_{he} = emission stream temperature after heat recovery, °F

Note that for the case of no heat recovery, T_{he} = T_e. The factor 1.1 accounts for an estimated heat loss of 10 percent in the incinerator. Supplementary heat requirements are based on maximum emission flow rate, and hence will lead to a conservative design. In contrast to thermal incineration, there is no minimum supplementary heat requirement specified for catalytic incineration since no fuel is needed for flame stabilization. Depending on the VOC concentration, emission stream temperature, and level of heat recovery, supplementary heat requirements may be zero when heat recovery is practiced.

A plot of Equation 4.3-3 is shown in Figure 4-7 where the ratio of H_f/Q_e is plotted against the emission stream heat content for two levels of heat recovery (zero and 50 percent). As an alternative to Equation 4.3-3, use Figure 4.7 directly to determine H_f. The figure is based on the following assumptions: (1) moisture content of the emission stream (M_e) is 2 percent, (2) emission stream temperature (T_e) is 100°F, (3) preheat temperature of the emission stream (T_{he}) is based on 50 percent heat recovery in the heat exchanger, and (4) net heating value of supplementary fuel (natural gas) is 882 Btu/scf.

Figure 4-7. Supplementary heat requirement vs. emission stream heat content (dilute stream/no combustion air).

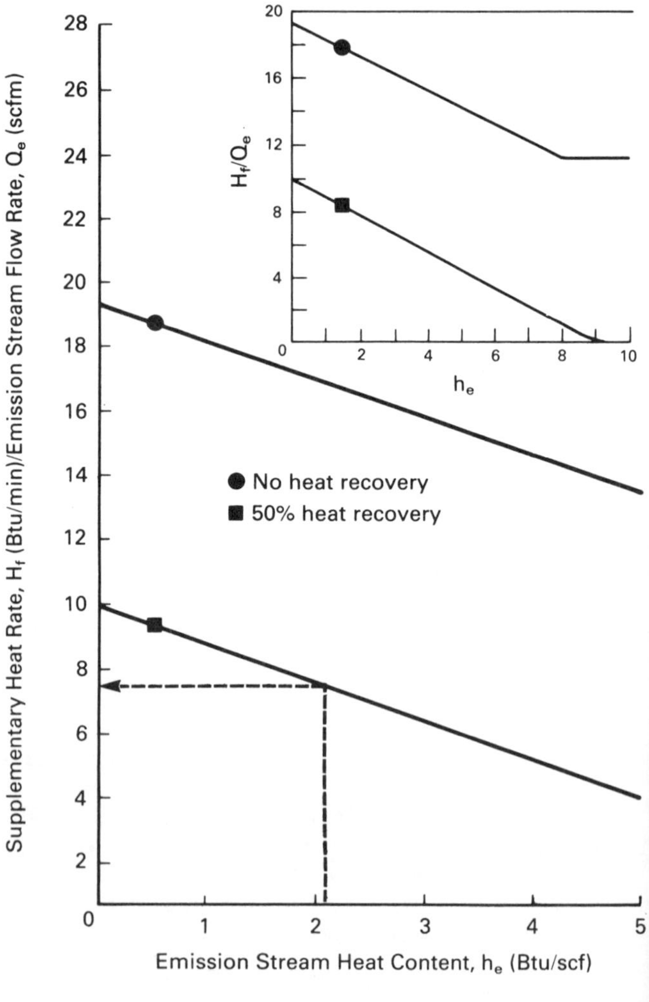

> **Example Case**
>
> *Using Equations 4.3-1, -2, and -3*:
>
> Since the emission stream is dilute (h_e = 2.1 Btu/scf) and has an oxygen concentration greater than 16% (O_2 = 20.6%), these equations are applicable.
>
> a. Determine if T_{co} falls in the range 1,000° - 1,200°F:
>
> T_{ci} = 600°F
> h_e = 2.1 Btu/scf (input data)
> T_{co} = 600 + (50 x 2.1) = 705°F
>
> Since T_{co} is less than 1,000°F, use Equation 4.3-2 to calculate an appropriate value for T_{ci}:
>
> T_{ci} = 1,000 - (50 x 2.1) = 895°F
>
> b. Determine H_f (assume recuperative heat recovery will be employed):
>
> Q_e = 20,000 scfm
> M_e = 2%
> T_r = 70°F
> T_{he} = 550°F (based on heat recovery of 50%)
> Cp_{air} = 0.0187 Btu/scf-°F for the interval 70° - 895°F (reference 3)
> Cp_{air} = 0.0183 Btu/scf-°F for the interval 70° - 550°F (reference 3)
>
> $$H_f = 1.1 \times 882 \times 20{,}000 \times 1.004 \left[\frac{(15.43 - 8.78)}{(882 - 21.60)}\right]$$
>
> H_f = 150,500 Btu/min
>
> *Using Figure 4-7*:
>
> For h_e = 2.1 Btu/scf and using the curve for 50% heat recovery, H_f/Q_e from the figure is about 7.5 Btu/min/scfm. Multiplying 7.5 by Q_e, (7.5 x 20,000), yields an approximate value of 150,000 Btu/min for H_f.

Figure 4-8. Supplementary heat requirement vs. emission stream heat content (no oxygen in emission stream/maximum combustion air).

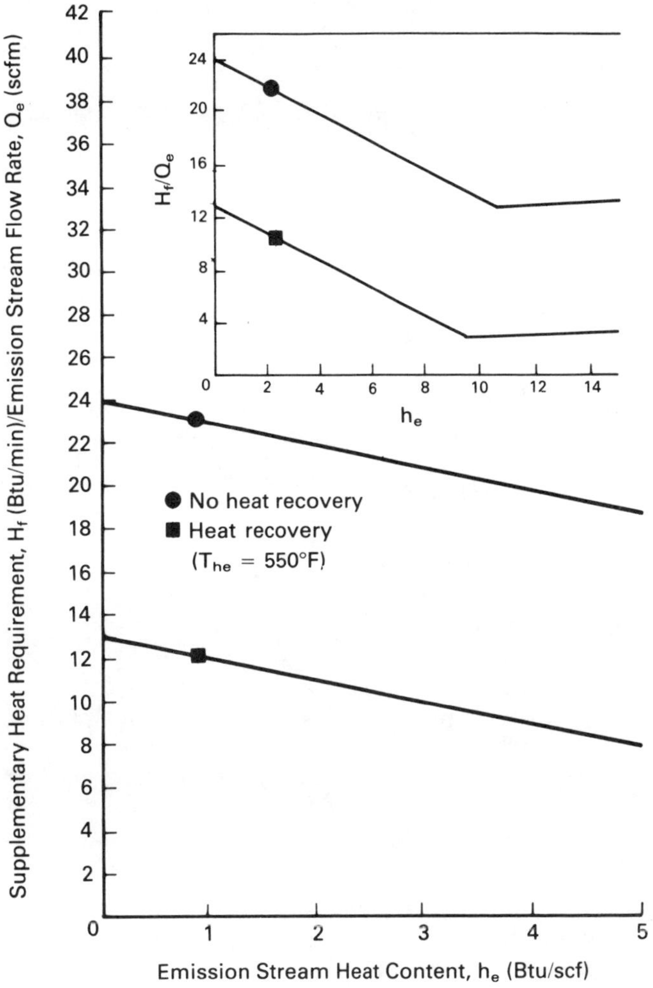

For emission streams that do not contain sufficient quantities of oxygen to satisfy combustion air requirements (e.g., process emissions), refer to Figure 4-8 which shows a plot of H_f/Q_e versus h_e for two levels of heat recovery (i.e., no heat recovery and where T_{he} is 550°F). In this figure, the oxygen content of the emission stream (O_2) is assumed as zero; this corresponds to maximum combustion air requirements. The emission stream moisture content (M_e), emission stream temperature (T_e) and the fuel heating value (h_f) are as specified for Figure 4-7. The preheat temperature of the emission stream (T_{he}) is 550°F for the heat recovery case. If O_2 for a particular emission stream is between 0 and 16 percent, use Figure 4-8.

4.3.4.2 Flow Rate of Combined Gas Stream Entering the Catalyst Bed

In order to calculate the quantity of catalyst required, the flow rate of the combined gas stream (emission stream + supplementary fuel combustion products) at the inlet to the catalyst bed has to be determined. Use the following equation:

$$Q_{com} = Q_e + Q_f + Q_c \qquad (4.3\text{-}4)$$

where:

Q_{com} = flow rate of the combined gas stream, scfm
Q_f = natural gas flow rate, scfm
Q_c = combustion air requirement, scfm

Calculate Q_f from the following expression:

$$Q_f = H_f/h_f \quad (4.3\text{-}5)$$

As indicated earlier, Q_c will typically be zero for dilute emission streams with oxygen contents (O_2) greater than 16 percent. If O_2 is less than 16 percent, then:

$$Q_c = (0.01\ H_e + 9.4\ Q_f)\ (1 + 0.01\ ex) - 0.476\ O_2Q_e$$

where:

H_e = heat generated due to the combustion of hydrocarbons in the emission stream

$H_e = Q_eh_e$

Example Case

Using Equation 4.3-4:

Q_e = 20,000 scfm

Q_c = 0 (since O_2 is greater than 16%)

Q_{com} = 20,000 + 170 + 0

Q_{com} = 20,170 scfm

Using Equation 4.3-5:

H_f = 150,500 Btu/min

h_f = 882 Btu/scf

Q_f = 170 scfm

4.3.4.3 Flow Rate of Flue Gas Leaving the Catalyst Bed

In order to determine costs for incinerators and to size a heat exchanger for preheating the emission stream, the flow rate of flue gas leaving the catalyst bed must be determined.

Assume that the flow rate of the combined gas stream entering the catalyst bed is approximately equal to the flow rate of the flue gas leaving the catalyst bed at standard conditions. The volume change across the catalyst bed due to the combustion of the VOC in the mixed gas stream is small, especially when dilute emission streams are treated. Therefore,

$$Q_{fg} = Q_{com}$$

where:

Q_{fg} = flow rate of the flue gas leaving the catalyst bed, scfm

While figuring costs, assume that catalytic incinerators are designed for a minimum Q_{fg} of 500 scfm. Therefore, if Q_{fg} is less than 500 scfm, define Q_{fg} as 500 scfm.

In order to determine operating costs, the flue gas flow rate (Q_{fg}) has to be expressed at actual conditions. Use the following equation to convert Q_{fg} from "scfm" to "acfm":

$$Q_{fg,a} = Q_{fg}[T_{co} + 460)/530] \quad (4.3\text{-}6)$$

where:

$Q_{fg,a}$ is the flue gas flow rate at actual conditions (acfm)

Example Case

Using Equation 4.3-6:

Q_{fg} = Q_{com} = 20,170 scfm

T_{co} = 1,000°F

$Q_{fg,a}$ = [20,170 (1,000 + 460)/530]

$Q_{fg,a}$ = 55,600 acfm

4.3.5 Catalyst Bed Requirement

The total volume of catalyst required for a given destruction efficiency is determined from the design space velocity as follows:

$$V_{bed} = 60\ Q_{com}/SV \quad (4.3\text{-}7)$$

where:

V_{bed} = volume of catalyst bed required, ft^3

Example Case

Using Equation 4.3-7:

Q_{com} = 20,170 scfm

SV = 30,000 hr^{-1} (Table 4-3)

V_{bed} = 60 x 20,170/30,000

V_{bed} = 40 ft^3

4.3.6 Heat Exchanger Size (for Systems with Recuperative Heat Exchange Only)

To determine the size of the heat exchanger required for preheating the emission stream to $T_{he,}$ use the following expression:

$$A = [60\ Q_e(1 + 0.002\ M_e)\ Cp_{air}\ (T_{he} - T_e)]/U\Delta T_{LM} \quad (4.3\text{-}8)$$

where:

A = heat exchanger surface area, ft_2

U = overall heat transfer coefficient, Btu/hr-ft^2-°F

ΔT_{LM} = logarithmic mean temperature difference, °F

and:

$$\Delta T_{LM} = [(T_c - T_{he}) - (T_{he} - T_e)]\ \ln\ [(T_c - T_{he})/(T_{he} - T_e)]$$

For dilute emission streams that do not require additional combustion air, then ΔT_{LM} can be approximated by:

$$\Delta T_{LM} = T_c - T_{he}$$

For a recuperative heat exchanger where the heat transfer takes place between two gas streams, the overall heat transfer coefficient (U) ranges from 2 to 8 Btu/hr-ft°F, generally depending on the heat exchanger configuration and properties of the gas streams.

Example Case

Using Equation 4.3-8:

Q_e = 20,000 scfm
M_e = 2%
T_{he} = 550°F (based on heat recovery of 50%)
T_e = 120°F (input data)
Cp_{air} = 0.0183 Btu/scf-°F for the interval 120° - 550°F (reference 3)
U = 4 Btu/hr-ft^2-°F

Since the emission stream is dilute, calculate ΔT_{LM} as follows:

$\Delta T_{LM} = T_{co} - T_{he}$
T_{co} = 1,000°F
ΔT_{LM} = 1,000 - 550 = 450°F

The heat exchanger surface area from Equation 4.3-8 then becomes:

$$A = \left[\frac{60 \times 20{,}000 \times 1.004 \times 0.0183 \times (550\text{-}100)}{4 \times 450}\right]$$

A = 5,500 ft^2

Using Figure 4-9:
For all values of h_e, A/Q_e is about 275 x 10^{-3}. Thus, multiplying Q_e by A/Q_e, (20,000 x 0.275) yields 5,500 ft^2.

Alternatively, Figure 4-9 can be used to determine the heat exchanger size. In this figure, line (1) represents Equation 4.3-8 evaluated for dilute emission streams that do not require additional combustion air. The assumptions inherent in this case are the same as those stated for Figure 4-7. The overall heat transfer coefficient is assumed as 4 Btu/hr-ft^2-°F.

For emission streams that require additional combustion air, use the solid line in Figure 4-9 to obtain an estimate of the required heat exchanger size. As in Figure 4-8, the solid line is based on maximum combustion air requirements (i.e., no oxygen in the emission stream). If Figure 4-9 is not directly applicable in a particular situation, use the broken line in the figure to obtain a conservative estimate for A.

4.3.7 Evaluation of Permit Application

Compare the results from the calculations and the values supplied by the permit applicant using Table 4-4. The calculated values in the table are based on the example case.

If the calculated values for H_f, Q_c, Q_{com}, V_{bed}, and A differ from the reported values for these variables, the differences may be due to the assumptions involved in the calculations. If that is the case, the reviewer may wish to discuss the details of the proposed design with the permit applicant.

If the calculated values agree with the reported values, however, then the design and operation of the proposed catalytic incinerator system may be considered appropriate based on the assumptions used in this handbook.

Figure 4-9. Heat exchanger size vs. emission stream heat content.

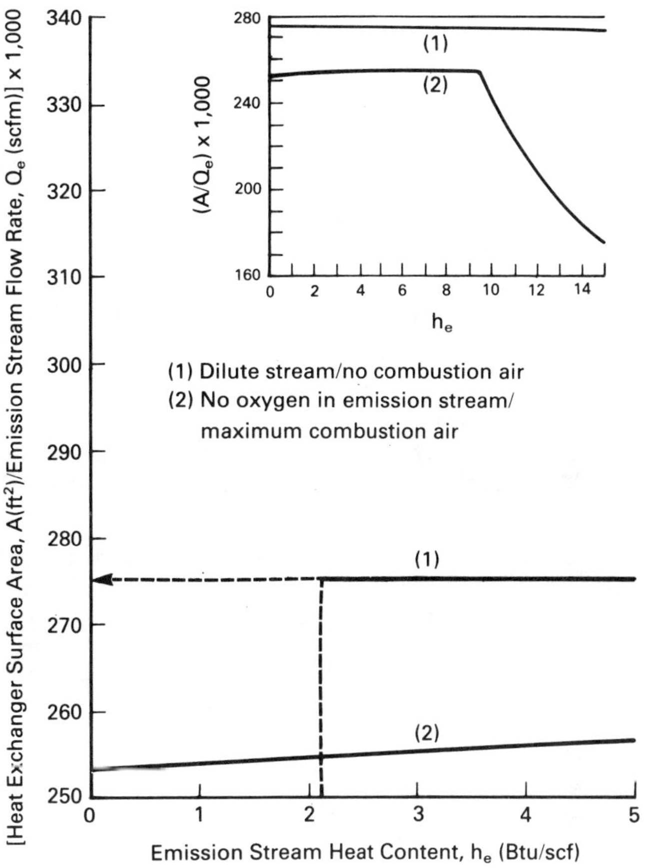

Table 4-4. Comparison of Calculated Values and Values Supplied by the Permit Applicant for Catalytic Incineration

	Calculated Value (Example Case)[a]	Reported Value
Supplementary heat requirement, H_f	150,500 Btu	...
Supplementary fuel flow rate, Q_f	170 scfm	...
Combustion air flow rate, Q_c	0	...
Combined gas stream flow rate, Q_{com}	20,170 scfm	...
Catalyst bed volume, V_{bed}	40 ft^3	...
Heat exchanger surface area (if recuperative heat recovery is used), A	5,500 ft^2	...

[a] Based on Emission Stream 2.

4.4 Flares

Flares use open flames for disposing of waste gases during normal operations and emergencies. They are typically applied when the heating value of the waste gases cannot be recovered economically because of intermittent or uncertain flow, or when process upsets occur. In some cases, flares are operated in conjunction with baseload gas recovery systems (e.g., condensers). Flares handle process upset and emergency gas releases that the baseload system is not designed to recover.

There are several types of flares, the most common of which are steam-assisted, air-assisted, and pressure head flares. Typical flare operations can be classified as "smokeless," "nonsmokeless," and "fired" or "endothermic." For smokeless operation, flares use outside momentum sources (usually steam or air) to provide efficient gas/air mixing and turbulence for complete combustion. Smokeless flaring is required for destruction of organics heavier than methane. Nonsmokeless operation is used for organic or other vapor streams which burn readily and do not produce smoke. Fired, or endothermic, flaring requires additional energy in order to ensure complete oxidation of the waste streams such as for sulfur tail gas and ammonia waste streams.

In general, flare performance depends on such factors as flare gas exit velocity, emission stream heating value, residence time in the combustion zone, waste gas/oxygen mixing, and flame temperature. Since steam-assisted smokeless flares are the most frequently used, they will be the focus of this discussion. A typical steam-assisted flare system is shown in Figure 4-10. First, process off-gases enter the flare through the collection header. When water or organic droplets are present, it may be necessary to pass the off-gases through a knock-out drum, since these droplets can create problems. Water droplets can extinguish the flame and organic droplets can result in burning particles.

Once the off-gases enter the flare stack, flame flashback can occur if the emission stream flow rate is too low. Flashback may be prevented, however, by passing the gas through a gas barrier, a water seal, or a stack seal. Purge gas is another option. At the flare tip, the emission stream is ignited by pilot burners. If conditions in the flame zone are optimum (oxygen availability, adequate residence time, etc.), the VOC in the emission stream may be completely burned (~100 percent efficiency). In some cases, it may be necessary to add supplementary fuel (natural gas) to the emission stream in order to achieve destruction efficiencies of 98 percent and greater if the net heating value of the emission stream is less than 300 Btu/scf.(9,10)

Typically, existing flare systems will be used to control HAP emission streams. Therefore, the fol-

Figure 4-10. A typical steam-assisted flare system.

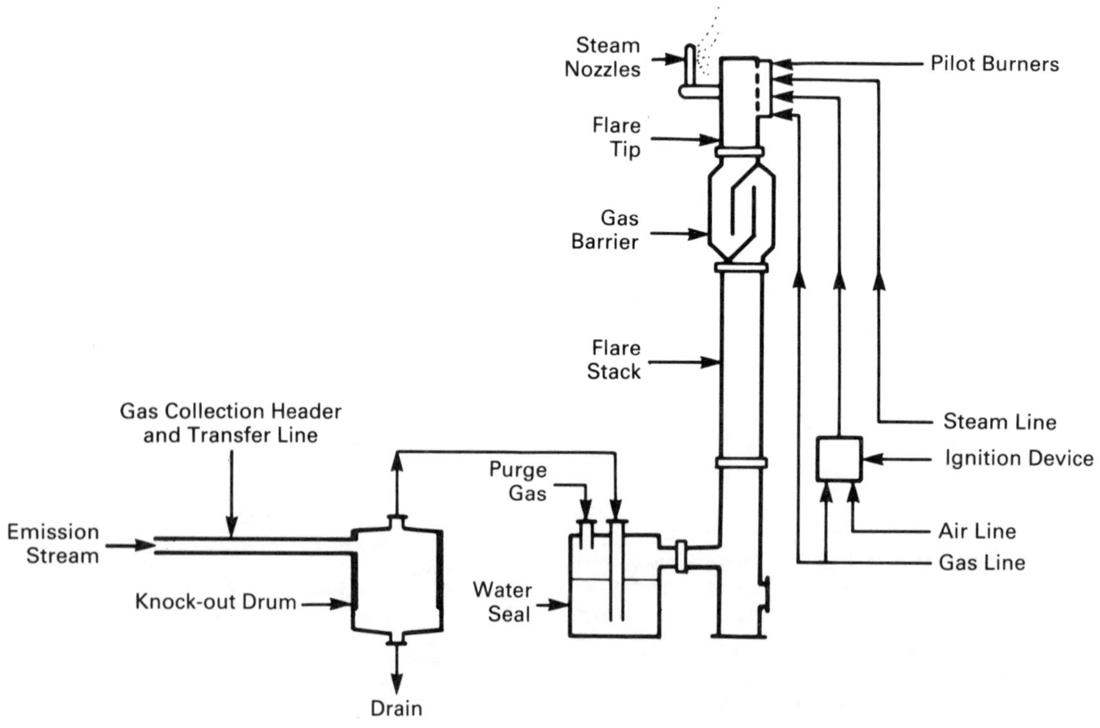

lowing sections describe how to evaluate the destruction efficiency of an existing flare system under expected flow conditions (e.g., continuous, start-up, shut-down, etc.). The discussion will be based on the recent regulatory requirements of 98 percent destruction efficiency for flares.(9) The calculation procedure will be illustrated for Emission Stream 3 described in Chapter 3 using a steam-assisted flare system. Note that flares often serve more than one process unit and the total flow rate to the flare needs to be determined before the following calculation procedure can be applied.

4.4.1 Data Required

The data necessary to perform the calculations consist of HAP emission stream characteristics previously compiled on the HAP Emission Stream Data Form, flare dimensions, and the required HAP control as determined by the applicable regulations.

> **Example Case**
>
> Expected emission stream flowrate,
> Q_e = 30,000 scfm
> Emission stream temperature, T_e = 100°F
> Heat content, h_e = 180 Btu/scf
> Mean molecular weight of emission stream,
> MW_e = 33.5 lb/lb-mole
> Flare tip diameter, D_{tip} = 48 in
>
> Based on the control requirements for the emission stream:
> Required destruction efficiency, DE = 98%

In the case of a permit review, the data outlined below should be supplied by the applicant. The calculations in this section will then be used to check the applicant's values. Worksheets are provided in Appendix C.5.

Flare system variables at standard conditions (70°F, 1 atm):

Flare tip diameter, D_{tip}, in
Expected emission stream flowrate, Q_e, scfm
Emission stream heat content, h_e, Btu/scf
Temperature of emission stream, T_e, °F
Mean molecular weight of emission stream, MW_e, lb/lb-mole
Steam flowrate, Q_s, lb/min
Flare gas exit velocity, U_{flg}, ft/sec
Supplementary fuel flow rate,[a] Q_f, scfm
Supplementary fuel heat content,[a] h_f, Btu/scf
Temperature of flare gas,[b] T_{flg}, °F
Flare gas flow rate,[b] Q_{flg}, scfm
Flare gas heat content,[b] h_{flg}, Btu/scf

[a] This information is needed if the emission stream heat content is less than 300 Btu/scf.
[b] If no auxiliary fuel is added, the value for this variable will be the same as that for the emission stream.
[c] For unassisted flares, the lower limit is 200 Btu/scf.

4.4.2 Determination of Flare Operating Variables

Based on studies conducted by EPA, relief gases having heating values less than 300 Btu/scf are not assured of achieving 98 percent destruction efficiency when they are flared in steam- or air-assisted flares.[c] Therefore, the first step in the evaluation procedure is to check the heat content of the emission stream and determine if additional fuel is needed.

In a permit review case, if the heat content of the emission stream is less than 300 Btu/scf and no supplementary fuel has been added, then the application is considered unacceptable. The reviewer may then wish to follow the calculations described below. If the reported value for the emission stream heat content is above 300 Btu/scf, however, then the reviewer should skip to Section 4.4.2.3.

4.4.2.1 Supplementary Fuel Requirements

If the emission stream heat content is less than the 300 Btu/scf required to achieve a destruction level of 98 percent, it is assumed that natural gas will be added to the emission stream to bring its heat content to 300 Btu/scf. Calculate the required natural gas requirements using the following equation:

$$Q_f = [(300 - h_e)Q_e]/582 \qquad (4.4\text{-}1)$$

where:
Q_e = emission stream flow rate, scfm
Q_f = natural gas flow rate, scfm
h_e = emission stream heat content, Btu/scf

(See Appendix B.7, reference 8, for details of all the equations used in Section 4.4.) If the emission stream heat content is greater than or equal to 300 Btu/scf, then Q_f = 0.

> **Example Case**
>
> *Using Equation 4.4-1*:
> Since h_e is less than 300 Btu/scf, supplementary fuel is needed.
> h_e = 180 Btu/scf
> Q_e = 30,000 scfm
> Q_f = (300 - 180)(30,000)/582
> Q_f = 6,200 scfm

4.4.2.2 Flare Gas Flow Rate and Heat Content

The flare gas flow rate is determined from the flow rates of the emission stream and natural gas using the following equation:

$$Q_{flg} = Q_e + Q_f \qquad (4.4\text{-}2)$$

where:
Q_{flg} = flare gas flow rate, scfm

Note that if Q_f = 0, then Q_{flg} = Q_e.

The heat content of the flare gas (h_{flg}) is dependent on whether supplementary fuel is added to the emission stream. When h_e is greater than or equal to 300 Btu/scf, then $h_{flg} = h_e$. If h_e is less than 300 Btu/scf, since supplementary fuel is added to increase h_e to 300 Btu/scf, $h_{flg} = 300$ Btu/scf.

Example Case

Using Equation 4.4-2:

Q_e = 30,000 scfm
Q_f = 6,200 scfm
Q_{flg} = 36,200 scfm
Since h_e = 180 Btu/scf, h_{flg} = 300 Btu/scf.

4.4.2.3 Flare Gas Exit Velocity

The flare gas exit velocity values presented in Table 4-5 to achieve at least 98 percent destruction efficiency in a steam-assisted flare system are based on studies conducted by EPA.(9) Flare gas exit velocities are expressed as a function of flare gas heat content. Determine the maximum allowable exit velocity using the equation presented in Table 4-5.

Table 4-5. Flare Gas Exit Velocities (9)

Flare Gas Heat Content[a] h_{flg} (Btu/scf)	Maximum Exit Velocity U_{max} (ft/sec)
$h_{flg} < 300$	–[b]
$300 \leq h_{flg} < 1{,}000$	$3.28\ [10^{(0.00118\ h_{flg}\ + 0.908)}]$
$h_{flg} \geq 1{,}000$	400

[a] If no supplementary fuel is used, $h_{flg} = h_e$.
[b] Based on studies conducted by EPA, waste gases having heating values less than 300 Btu/scf are not assured of achieving 98% destruction efficiency when they are flared in steam-assisted flares.(4)

Example Case

Since h_{flg} = 300 Btu/scf, use the equation in Table 4-5 to calculate U_{max}:

$$U_{max} = 3.28\ [10^{(0.00118\ h_{flg}\ +\ 0.908)}]$$
$$= 3.28\ [10^{(0.00118\ \times\ 300\ +\ 0.908)}]$$
$$U_{max} = 60 \text{ ft/sec}$$

From the emission stream data (expected flow rate, temperature) and information on flare diameter, calculate the flare gas exit velocity (U_{flg}); compare this value with U_{max}. Use the following equation to calculate U_{flg}:

$$U_{flg} = (Q_{flg,a}/60)\ [4/\pi(D_{tip}/12)^2]$$
$$= (3.06\ Q_{flg,a})/(D_{tip})^2 \qquad (4.4\text{-}3)$$

where:

U_{flg} = exit velocity of flare gas, ft/sec
$Q_{flg,a}$ = flare gas flow rate at actual conditions, acfm
D_{tip} = flare tip diameter, in

Use the following expression to calculate $Q_{flg,a}$:

$$Q_{flg,a} = [Q_{flg}\ (T_{flg} + 460)]/530 \qquad (4.4\text{-}4)$$

If U_{flg} is less than U_{max}, then the 98 percent destruction level can be achieved. However, if U_{flg} exceeds U_{max}, this destruction efficiency level cannot be achieved. This indicates that the existing flare diameter is too small to accommodate lower exit velocities for the emission stream under consideration. Note that at very low flare gas exit velocities, flame instability may occur. The minimum flare gas exit velocity for a stable flame is assumed as 0.03 ft/sec in this handbook.(1) Thus, if U_{flg} is below 0.03 ft/sec, the desired destruction efficiency may not be achieved. In summary, U_{flg} should fall in the range 0.03 ft/sec and U_{max} for a 98 percent destruction efficiency level.

In a permit review case, if U_{flg} exceeds U_{max}, then the application is not acceptable. If U_{flg} is below U_{max} and exceeds 0.03 ft/sec, then the proposed design is considered acceptable and the reviewer may proceed with the calculations.

Example Case

Using Equations 4.4-3 and -4:

Q_{flg} = 36,200 scfm
T_{flg} = 95°F
$Q_{flg,a}$ = [36,200 (95 + 460)]/530
$Q_{flg,a}$ = 37,900 acfm

D_{tip} = 48 in
U_{flg} = $(3.06 \times 37{,}900)/(48)^2$
U_{flg} = 50 ft/sec

Since 0.03 ft/sec < U_{flg} = 50 ft/sec < U_{max} = 60 ft/sec, the required level of 98% destruction efficiency can be achieved under these conditions.

4.4.2.4 Steam Requirements

Steam requirements for steam-assisted flare operation depend on the composition of the flare gas and the flare-tip design. Typical values range from 0.15 to 0.50 lb steam/lb flare gas. In this handbook, the amount of steam required for 98 percent destruction efficiency is assumed as 0.4 lb steam/lb flare gas.(2) Use the following equation to determine steam requirements:

$$Q_s = 1.03 \times 10^{-3} \times Q_{flg} \times MW_{flg} \qquad (4.4\text{-}5)$$

where:

Q_s = steam requirement, lb/min
MW_{flg} = $(Q_f \times 16.7 + Q_e MW_e)/Q_{flg}$

Example Case

Using Equation 4.4-5:

Q_{flg} = 36,200 scfm
MW_{flg} = 30.6 lb/lb-mole
Q_s = $1.03 \times 10^{-3} \times 36{,}200 \times 30.6$
Q_s = 1,140 lb/min

4.4.3 Evaluation of Permit Application

Compare the results from the calculated and reported values using Table 4-6. If the calculated values of Q_f, U_{flg}, Q_{flg}, and Q_s are different from the reported values for these variables, the differences may be due to the assumptions (e.g., steam to flare gas ratios, etc.) involved in the calculations. In such a case, the reviewer may wish to discuss the details of the proposed system with the permit applicant.

If the calculated values agree with the reported values, then the operation of the proposed flare system may be considered appropriate based on the assumptions made in this handbook.

Table 4-6. Comparison of Calculated Values and Values Supplied by the Permit Applicant for Flares

	Calculated Value (Example Case)[a]	Reported Value
Supplementary fuel flow rate, Q_f	6,200 scfm	...
Flare gas exit velocity, U_{flg}	50 ft/sec	...
Flare gas flow rate, Q_{flg}	36,200 scfm	...
Steam flow rate, Q_s	1,140 lb/min	...

[a] Based on Emission Stream 3.

4.5 Boilers/Process Heaters

The application of boilers and/or process heaters as emission control devices is very site-specific (see Section 3.2.1.4). The level of detail required in the calculations for sizing such devices is beyond the scope of this handbook and thus is not presented.

4.6 Carbon Adsorption

Adsorption is a surface phenomenon whereby hydrocarbons and other compounds are selectively adsorbed on the surface of such materials as activated carbon, silica gel, or alumina. Activated carbon is the most widely used adsorbent. The adsorption capacity of an adsorbent for a given VOC is often represented by adsorption isotherms that relate the amount of VOC adsorbed (adsorbate) to the equilibrium pressure (or concentration) at constant temperature (see Figure 4-11). Typically, the adsorption capacity increases with the molecular weight of the VOC being adsorbed. In addition, unsaturated compounds are generally more completely adsorbed than saturated compounds, and cyclical compounds are more easily adsorbed than linearly structured materials. Also, the adsorption capacity is enhanced by lower operating temperatures and higher concentrations. VOC's characterized by low vapor pressures are more easily adsorbed than those with high vapor pressures.

Carbon adsorption is used for pollution control and/or for solvent recovery in several industries. It is usually a batch operation, involving multiple beds. The two main steps in the adsorption operation include adsorption and regeneration, usually performed cyclically. For control of continuous emission streams, at least one bed remains on line in the adsorption mode while the other is being regenerated.

Figure 4-11. Adsorption isotherms for toluene/activated carbon system.(11)

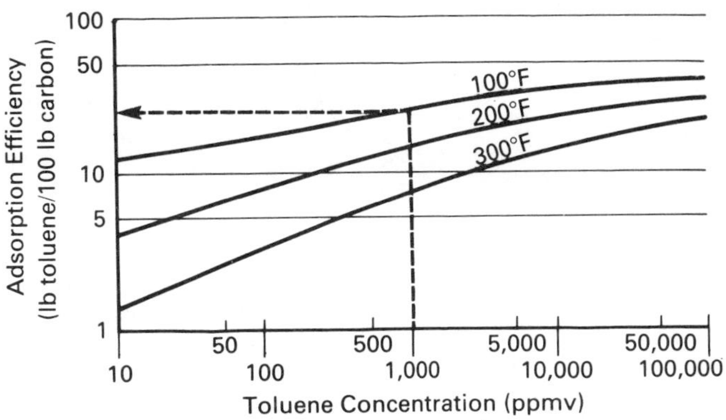

A typical batch operation (see Figure 4-12) can be described as follows: The VOC-laden waste gas is passed through the carbon bed where the VOC's are adsorbed on the bed surface. As the adsorption capacity of the bed is approached, traces of VOC's appear in the exit stream, indicating that the breakthrough point of the bed has been attained. The emission stream is then directed to a parallel bed containing regenerated adsorbent, and the process continued. Concurrently, the saturated bed is regenerated by the passage of hot inert gases, low-pressure steam, or a combination of vacuum and hot gas. Since adsorption is a reversible process, by supplying heat (equivalent to the amount of heat released during adsorption), the VOC's on the bed can be desorbed. A "heel" is always left on the bed because complete desorption is technically difficult to achieve and economically impractical.

During the last part of the steam regeneration cycle, the hot bed saturated with water vapor is dried and cooled, usually with air. When steam is used as the regenerant, the desorbed VOC's in the steam effluent are typically condensed. Then the VOC's are recovered either by simple decantation, in the case of water-insoluble materials, or by distillation, in the case of water-soluble materials. If high purity is required for the recovered VOC's, complex distillation systems may be necessary, especially in cases where the VOC's consist of mixtures of solvents.

In other designs, continuous adsorption can be accomplished by fluidized bed adsorption. The fresh adsorbent flows down the adsorption section that consists of a series of fluidized trays. The emission

Figure 4-12. A typical fixed-bed carbon adsorption system.

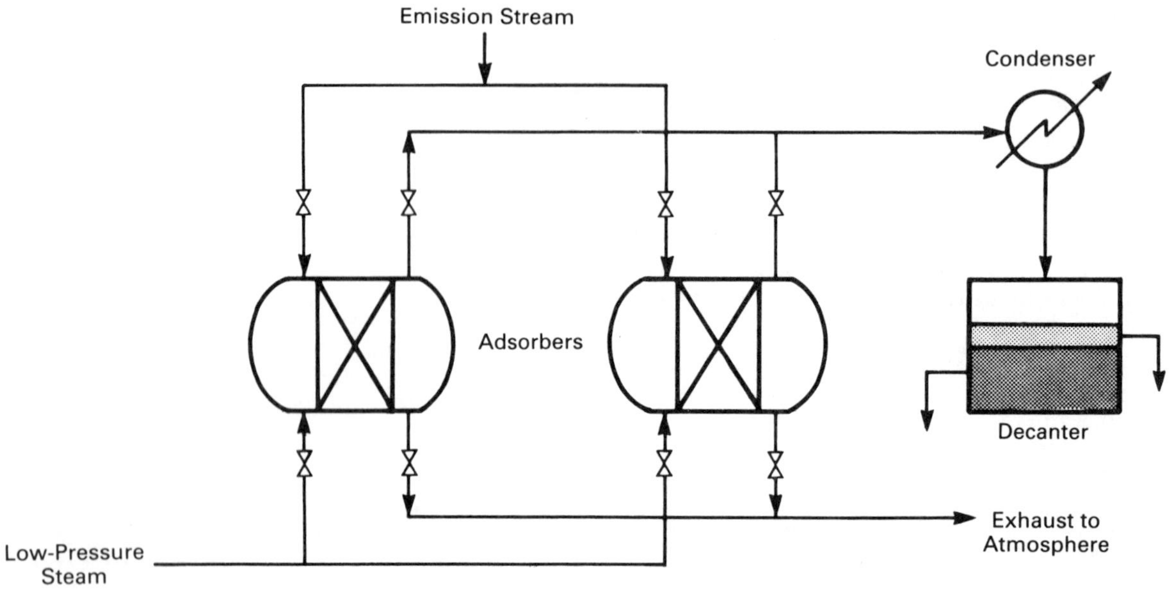

stream enters at the bottom of the adsorption section and flows upward. The VOC's are progressively adsorbed and the exit gas is discharged from the top stage. The saturated adsorbent is continuously removed from the bottom and transferred to the desorption section, where it is regenerated and returned to the adsorption system.

For a given emission stream, the performance of a carbon adsorber as a control device is affected by several variables including: (a) the adsorption capacity of the carbon for the specific VOC in question (as determined from the adsorption isotherm), (b) operating temperature, (c) adsorption and regeneration cycle time, (d) amount and type of regenerant, and (e) contaminants.

The discussion in the following sections will be based on a fixed-bed carbon adsorption system with two parallel beds. Regeneration of the beds will be carried out with low pressure steam. It is assumed that the desorbed VOC's and steam will be condensed and the bed will be dried and cooled with air. Another assumption is that the emission stream is free from liquid and/or solid particles that may potentially blind the carbon beds. Emission Stream 4 described in Chapter 3 will be used to illustrate the calculation procedure. Worksheets for calculations are provided in Appendix C.6.

4.6.1 Data Required

The data necessary to perform the calculations consist of HAP emission stream characteristics previously compiled on the HAP Emission Stream Data Form and the required HAP control as determined by the applicable regulations.

Example Case

Maximum flowrate, Q_e = 15,000 scfm
Temperature, T_e = 90°F
Relative humidity, R_{hum} = 40%
HAP = toluene
Maximum HAP content, HAP_e = 1,000 ppmv

Based on the control requirements for the emission stream:
Required removal efficiency, RE = 95%

If dilution air is added to the emission stream upon exit from process, the data that will be used in the calculations are the resulting characteristics after dilution.

In a permit review case for a carbon adsorber, the following data should be supplied by the applicant. The calculations in this section will later be used to check the applicant's values.

Carbon adsorber (fixed-bed) system variables at standard conditions (70°F, 1 atm):

Reported removal efficiency, $RE_{reported}$, %
HAP content, HAP_e, ppmv
Emission stream flow rate, Q_e, scfm
Adsorption capacity of carbon bed, AC, lb HAP/100 lb carbon
Number of beds, N
Amount of carbon required, C_{req}, lb
Cycle time for adsorption, θ_{ad}, hr
Cycle time for regeneration, θ_{reg}, hr
Emission stream velocity through the carbon bed, U_e, ft/min
Bed depth, Z_{bed}, ft
Bed diameter, D_{bed}, ft
Steam ratio, St, lb steam/lb carbon

4.6.2 Pretreatment of the Emission Stream

4.6.2.1 Cooling

Adsorption of VOC's is favored by lower temperatures. If the temperature of the emission stream is significantly higher than 100°F, a heat exchanger may be needed to cool the emission stream to 100°F. (See Appendix B.5, reference 8, for determining the size of a heat exchanger required for such applications.

> **Example Case**
> The temperature of the emission stream is 90°F, which is below 100°F. Therefore, cooling is not necessary.

4.6.2.2 Dehumidification

Since water vapor competes with the VOC's in the emission stream for adsorption sites on the carbon surface, emission stream humidity levels exceeding 50 percent (relative humidity) are not desirable. In this handbook, it is assumed that if the relative humidity level of the emission stream is above 50 percent, it will be reduced to 50 percent using additional equipment.

Dehumidification may be carried out by cooling and condensing the water vapor in the emission stream. A shell-and-tube type heat exchanger can be employed for this purpose. Refer to Section 4.8 where calculation procedures for sizing condensers are described.

Another alternative for dehumidification is adding dilution air to the emission stream if the dilution air humidity is significantly less than that of the emission stream. However, since this will increase the size of the adsorber system required, it may not be cost effective.

> **Example Case**
> Since the relative humidity of the emission stream is less than 50%, dehumidification is not necessary.

4.6.2.3 High VOC Concentrations

If flammable vapors are present in emission streams that are mixtures of VOC and air, the VOC content may be limited to below 25 percent of the LEL by insurance companies. In some cases, it can be increased to 40 to 50 percent of the LEL if proper monitoring and controls are used. In addition, since high bed temperatures may occur due to heat released during adsorption, high VOC concentrations may need to be reduced. The maximum practical inlet VOC concentration is usually about 10,000 ppmv. In this handbook, it is assumed that the VOC content will be limited to less than 25 percent of the LEL. See Table B.1-2 in Appendix B.1 for a list of LEL values for several compounds.

> **Example Case**
> The HAP concentration of the emission stream is 1,000 ppmv (toluene). This is below 25% of the LEL for toluene, which is 3,000 ppmv (see Table B.1-2).

4.6.3 Carbon Adsorption System Design Variables

Table 4-7 presents suggested values for the design variables of a carbon adsorber system to achieve a given outlet HAP concentration. If the emission limit requirement is expressed as removal efficiency, the outlet HAP concentration can be calculated from the required removal efficiency and the inlet HAP concentration.

Table 4-7. Carbon Adsorber System Design Variables (12,13)

Outlet Concentration[a] HAP_o (ppmv)	Adsorption Cycle Time μ_{ad}(hr)	Degeneration Cycle Time[b] μ_{reg} (hr)	Steam Requirement for Regeneration St (lb steam/ lb carbon)
70	2	2	0.3
10 - 12	2	2	1.0

[a] Emission stream exiting the carbon adsorber.
[b] Regeneration cycle also includes the time necessary for drying and cooling the bed.

For specific applications, other values may be appropriate depending on the emission stream characteristics and the type of carbon bed. For example, the adsorption capacity for a given carbon bed is dependent on several factors, including the type of VOC in the emission stream and the temperature and humidity levels. Typically, the adsorption capacity is determined from the adsorption isotherm of the compound under consideration. (See Appendix B.8, reference 8, or references 14 and 15 for adsorption isotherms for several compounds. Also see reference 8 or 16 for adsorption capacities at specific conditions for several compounds.)

Based on the required removal efficiency, determine the outlet HAP concentration using the following equation:

$$HAP_o = HAP_e\,(1 - 0.01\,RE) \qquad (4.6\text{-}1)$$

where:

HAP_o = HAP content of the emission stream exiting the adsorber, ppmv

RE = removal efficiency, %

Next, specify the appropriate values for θ_{ad}, θ_{req}, and St using Table 4-7.

> **Example Case**
> *Using Equation 4.6-1:*
> RE = 95 percent
> HAP_e = 1,000 ppmv
> HAP_o = 1,000 (1 - 0.95)
> HAP_o = 50 ppmv
>
> Assuming the conditions for HAP_o = 70 ppmv are approximately the same as those for HAP_o = 50 ppmv, from Table 4-7,
> θ_{ad} = 2 hrs
> θ_{reg} = 2 hrs
> St = 0.3 lb steam/lb carbon

4.6.4 Determination of Carbon Adsorber System Variables

4.6.4.1 Carbon Requirements

In sizing a carbon adsorber system, the quantity of carbon required is determined from the adsorption capacity of the carbon bed (based on the adsorption isotherm of the HAP in question) using the emission stream flow rate and HAP concentration. For a fixed-bed adsorption system with N parallel beds and a specified adsorption cycle θ_{ad}, the following equation can be used to calculate the carbon requirements:

$$C_{req} = 2\,[1.55 \times 10^{-5} N\, \theta_{ad}\, Q_e\, (HAP_e - HAP_o) MW_{HAP}/AC] \qquad (4.6\text{-}2)$$

where:
C_{req} = carbon requirement, lb carbon
N = number of carbon beds
θ_{ad} = adsorption cycle time, hr
Q_e = emission stream flow rate, scfm
MW_{HAP} = molecular weight of HAP, lb/lb-mole (for a mixture of HAPs, MW_{HAP} will be defined as mean molecular weight)
AC = adsorption capacity of the carbon bed, lb HAP/100 lb carbon

For design purposes, the carbon requirement is generally multiplied by a factor of 2 as indicated in Equation 4.6-2.(17) This safety factor is an allowance for build-up of a heel during regeneration (which results in a reduced capacity); fluctuations in emission stream characteristics (e.g., HAP concentration and composition, humidity, etc.). The value for AC is typically determined from the adsorption isotherm of the specific HAP/carbon system. If an isotherm for the HAP in question is not available, the isotherm for another compound of similar molecular weight and boiling point may be used as an approximation. See Table B.1-3 in Appendix B.1 for molecular weight and boiling point data for several compounds. For additional data, see reference 18. If no data are available, use a conservative value of 5 lbs HAP/100 lb carbon for AC.

As an alternative, Figure 4-13 can be used to determine carbon requirements. The figure is based on Equation 4.6-2 and evaluated at HAP_o = 10 and 70 ppmv for several inlet concentrations.

If the emission stream contains a mixture of HAP's, Equation 4.6-2 should be evaluated using appropriate adsorption isotherms for each component and then summed to determine C_{req}.

> **Example Case**
> *Using Equation 4.6-2:*
> Assume N = 2
> θ_{ad} = 2 hrs (from Table 4-7)
> Q_e = 15,000 scfm
> HAP_e = 1,000 ppmv
> HAP_o = 50 ppmv
> MW_{HAP} = $MW_{toluene}$ = 92 lb/lb mole
> AC = 20-25 lb toluene/100 lb carbon
>
> This value is estimated from Figure 4-11 where adsorption isotherms for toluene are plotted at different temperatures. To obtain a conservative estimate for C_{req}, assume AC = 20:
>
> C_{req} = [2 x 1.55 x 10^{-5} x 2 x 2 x 15,000 (1,000 - 50) 92/20]
> C_{req} = 8,100 lb
>
> *Using Figure 4-13:*
> For HAP_e = 1,000 ppmv, HAP_o = 50 ppmv, and AC = 20 lb, C_{req}/Q_e is about 0.55, assuming the curve for HAP_o = 70 ppmv is applicable. Thus,
>
> C_{req} = 0.55 x 15,000 = 8,250 lbs

4.6.4.2 Carbon Adsorber Size

The size of the adsorber is determined using a two step calculation. First, using the actual flow rate of the emission stream and its velocity, calculate the bed area. Typically, velocities used in industry range from 50 to 100 ft/min depending on the system pressure. At high velocities, the bed pressure drop becomes too high for standard blowers; at lower velocities, the bed becomes too large and expensive. A value of 100 ft/min is assumed in this discussion. Use the following equation to calculate the required bed area:

$$A_{bed} = Q_{e,a}/U_e \qquad (4.6\text{-}3)$$

where:
A_{bed} = bed area, ft^2
$Q_{e,a}$ = emission stream flow rate at actual conditions, acfm
U_e = emission stream velocity, ft/min

In this expression, $Q_{e,a}$ is determined as follows:

$$Q_{e,a} = Q_e\,[T_e + 460)/530] \qquad (4.6\text{-}4)$$

Figure 4-13. Carbon requirement vs. HAP inlet concentration.

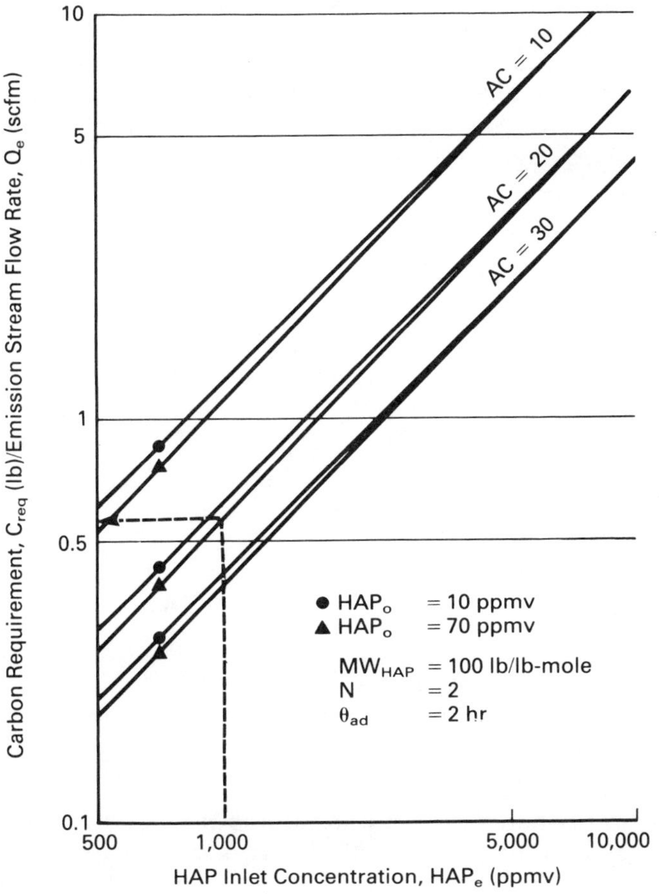

where Q_e and T_e are the flow rate and temperature of the emission stream. From the bed area, calculate the bed diameter assuming a circular vessel; use the following expression:

$$D_{bed} = 2\,[A_{bed}/\pi]^{0.5} = 1.13\,(A_{bed})^{0.5} \qquad (4.6\text{-}5)$$

where:

D_{bed} = bed diameter, ft

To calculate the bed depth, determine the volume occupied by carbon in each bed. Assume a carbon bed density of ρ_{bed}, and use the following equation to calculate the volume of carbon (per bed):

$$V_{carbon} = (C_{req}/N)/\rho_{bed} \qquad (4.6\text{-}6)$$

Having calculated V_{carbon}, the bed depth can be determined as follows:

$$Z_{bed} = V_{carbon}/A_{bed} \qquad (4.6\text{-}7)$$

where Z_{bed} is the bed depth, ft. Hence, the required adsorber size for an adsorption cycle time θ_{ad} for obtaining RE percent removal efficiency for an emission stream with flowrate Q_e is: D_e ft (diameter) by Z_{bed} ft straight side (minimum). Note that in cases where large flows (> 20,000 scfm) of off-gases are handled, three or more parallel beds may be used, reducing the bed size. The cycle times for adsorption and regeneration will change accordingly.

Example Case

Using Equations 4.6-3, -4, -5, -6, and -7:

T_e = 90°F
Q_e = 15,000 scfm
$Q_{e,a}$ = 15,000[(90 + 460)/530]
$Q_{e,a}$ = 15,565 acfm

U_e = 100 ft/min
A_{bed} = 15,565/100 = 155.7 ft^2
D_{bed} = 1.13 x $(155.7)^{0.5}$ = 14 ft

Assume ρ_{bed} = 30 lb/ft^3
V_{carbon} = (8,100/2)/30 = 135 ft^3
Z_{bed} = 135/155.7 = 0.88 = 1 ft

4.6.4.3 Steam Required for Regeneration

Carbon beds may be regenerated by various means; the most common regenerant used is steam. In this handbook, regeneration with steam is followed by condensation. The quantity of steam required for regeneration depends on the required removal efficiency (or outlet concentration) and on how much material is to be desorbed from the bed. A certain amount of steam is required to raise the bed to its regeneration temperature and provide the heat of desorption. The major portion of the steam flow, about 60 to 70 percent, acts as a carrier gas for the desorbed VOC's. It is not cost-effective to achieve complete desorption; acceptable working capacities of adsorption can be obtained without consuming large quantities of steam. For solvent recovery systems a requirement of 0.25 to 0.35 lb steam/lb carbon has usually been specified. For applications where VOC concentrations are low (e.g., odor control), steam usage ratios are higher.

In this handbook, it is assumed that with a steam ratio of 0.3 lb steam/lb carbon, a HAP outlet concentration of 70 ppmv can be achieved after regeneration, and with a ratio of 1 lb steam/lb carbon, a HAP outlet concentration of 10-12 ppmv can be achieved (Table 4-7).(11) The regeneration cycle time, θ_{reg}, is dependent on the time required to regenerate, dry, and cool the bed. The flow rate of the steam used for regeneration can be determined using the following expression:

$$Q_s = [St(C_{req})/(\theta_{reg} - \theta_{dry\text{-}cool})]/60 \qquad (4.6\text{-}8)$$

where:

Q_s = steam flowrate, lb/min
$\theta_{dry\text{-}cool}$ = cycle time for drying and cooling the bed, hr

Typically, cooling and drying the bed with air can be carried out in about 15 minutes. Figure 4-14 can also be used to estimate steam requirements. This figure is based on θ_{reg} = 2 hrs and $\theta_{dry\text{-}cool}$ = 0.25 hrs.

Steam flow rates based on cross-sectional area of the bed (Q_s/A_{bed}) are generally limited to less than 4 lb steam/min-ft^2 to prevent the carbon from being fluidized in the bed. If Q_s/A_{bed} exceeds 4, the regeneration cycle time or the steam ratio may need to be modified.

Example Case

Using Equation 4.6-8:
RE = 95%
HAP_o = 50 ppmv
St = 0.3 lb steam/lb carbon (Table 4-7)
θ_{reg} = 2 hrs (Table 4-7)

Assuming $\theta_{dry\text{-}cool}$ = 0.25 hrs:
Q_s = [0.3 (8,100)/(2 - 0.25)]/60
Q_s = 23 lb/min
Q_s/A_{bed} = 23/155.7 = 0.15 lb steam/min-ft^2

Since Q_s/A_{bed} is less than 4 lb steam/min ft^2, fluidization in the carbon bed is not expected.

Using Figure 4-14:
At C_{req} = 8,100 lb and St = 0.3,
Q_s = 20 lb/min

Figure 4-14. Steam requirement vs. carbon requirement.

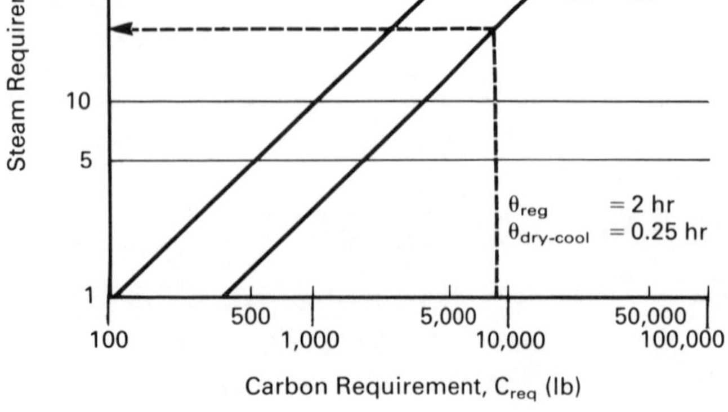

4.6.4.4 Condenser

The steam used for regenerating the carbon bed containing the desorbed VOC's is typically condensed. The heat duty of the condenser is based on the amount of steam required to regenerate the bed; the amount of heat absorbed by the bed and later removed by the drying and cooling air can be considered negligible.

Use the following expression to calculate the condenser size:

$$A_{con} = H_{load}/U\Delta T_{LM} \qquad (4.6\text{-}9)$$

where:
A_{con} = condenser surface area, ft^2
H_{load} = condenser heat load, Btu/hr
U = overall heat transfer coefficient, Btu/hr-ft^2-°F
ΔT_{LM} = logarithmic mean temperature difference, °F

and:

$$\Delta T_{LM} = \left[\frac{(T_{sti} - T_{wo}) - (T_{sto} - T_{wi})}{\ln\left[(T_{sti} - T_{wo})/(T_{sto} - T_{wi})\right]}\right]$$

where:
T_{sti} = steam inlet temperature, °F
T_{sto} = condensed steam outlet temperature, °F
T_{wo} = cooling water outlet temperature, °F
T_{wi} = cooling water inlet temperature, °F

In this handbook, it is assumed that following regeneration, steam will be condensed and subcooled to 100°F with cooling water. Note that Equation 4.6-9 is an approximate expression; condensation and subcooling processes are combined and average values are used for U and ΔT_{LM}. To calculate H_{load}, use the following equation:

$$H_{load} = 1.1 \times 60 \times Q_s\,[\lambda + \underline{C}p_w\,(T_{sti} - T_{sto})] \qquad (4.6\text{-}10)$$

where:
λ = latent heat of vaporization, Btu/lb
$\underline{C}p_w$ = average specific heat of water, Btu/lb-°F

In this expression, the condenser heat load is oversized by 10 percent. In the following calculations, steam available for regeneration is at atmospheric pressure and 212°F. The latent heat of vaporization of steam at these conditions is 970 Btu/lb.(19) The temperature rise of the cooling water available at 80°F is 50°F. The overall heat transfer coefficient is assumed as 150 Btu/hr-ft^2-°F.(12)

To determine cooling water requirements, use the following equation:

$$Q_{cool,w} = H_{load}/[\underline{C}p_w\,(T_{wo} - T_{wi})] \qquad (4.6\text{-}11)$$

where:
$Q_{cool,w}$ = cooling water flow rate, lb/hr
$Q_{cool,w}$ can be expressed in terms of gal/min as follows:

$$Q_w = Q_{cool,w} \times [(1/60) \times (1/62.43) \times 7.48] \\ = 0.002 \times Q_{cool,w} \quad (4.6\text{-}12)$$

where the factor 62.43 is the density of water and the factor 7.48 is used for converting from "ft^3" to "gal" basis.

Example Case

Using Equation 4.6-10:

Q_s = 23 lb/min
λ = 970 Btu/lb (reference 19)
Cp_w = 1 Btu/lb-°F (reference 19)
T_{sti} = 212°F
T_{sto} = 100°F
H_{load} = 1.1 x 60 x 23 [970 + 1 x (212 - 100)]
H_{load} = 1,642,500 Btu/hr

Using Equation 4.6-9:

U = 150 Btu/hr-ft^2-°F

$$\Delta T_{LM} = \left[\frac{(212\text{-}130) - (100\text{-}80)}{\ln[(212)130)/(100)80)]}\right]$$

ΔT_{LM} = 44°F

A_{con} = 1,642,500/(150 x 44)
A_{con} = 250 ft^2

Using Equations 4.6-11 and -12:

ΔT = 50°F
Q_w = 0.002 $Q_{cool,w}$
Q_w = 0.002 [1,642,500/(1x50)]
Q_w = 66 gal/min

4.6.4.5 Recovered Product

To calculate costs, the quantity of recovered product that can be sold and/or recycled to the process has to be calculated. Use the following equation:

$$Q_{rec} = 60 \times [Q_e \times (HAP_e \times 10^{-6})(1/387) \\ (0.01RE)\ MW_{HAP}] \quad (4.6\text{-}13)$$

$$Q_{rec} = 1.55 \times 10^{-9}\ Q_e \times HAP_e \times RE \times MW_{HAP}$$

where Q_{rec} is the quantity of recovered product, lb/hr. In this equation, the factor 387 is the volume (scf) occupied by 1 lb-mole of ideal gas at standard conditions (70°F and 1 atm).

Example Case

Using Equation 4.6-13:

Q_e = 15,000 scfm
HAP_e = 1,000 ppmv
RE = 95 %
MW_{HAP} = 92 lb/lb-mole
Q_{rec} = 1.55 x 10^{-9} x 15,000 x 1,000 x 95 x 92
Q_{rec} = 200 lb/hr

4.6.5 Evaluation of Permit Application

Compare the results from the calculated values and reported values using Table 4-8. If the calculated values of C_{req}, D_{bed}, Z_{bed}, Q_s, A_{con}, Q_w, and Q_{rec} are different from the reported values for these variables, the differences may be due to the assumptions involved in the calculations. Therefore, the reviewer may wish to discuss the details of the proposed design with the permit applicant.

Table 4-8. Comparison of Calculated Values and Values Supplied by the Permit Applicant for Carbon Adsorption

	Calculated Value (Example Case)[a]	Reported Value
Carbon requirement, C_{req}	8,100 lb	...
Bed diameter, D_{bed}	14 ft	...
Bed depth, Z_{bed}	1 ft	...
Steam rate, Q_s	23 lb/min	...
Condenser surface area, A_{con}	250 ft^2	...
Cooling water rate, Q_w	66 gal/min	...
Recovered product, Q_{rec}	200 lb/hr	...

[a] Based on Emission Stream 4.

If the calculated values agree with the reported values, then the design and operation of the proposed carbon adsorber system may be considered appropriate based on the assumptions made in this handbook.

4.7 Absorption

Absorption is an operation in which one or more components of a gas mixture are selectively transferred into a relatively nonvolatile liquid. Absorption of a gaseous component by a liquid occurs when the liquid contains less than the equilibrium concentration of the gaseous component. The difference between the actual concentration and the equilibrium concentration provides the driving force for absorption. The absorption rate depends on the physical properties of the gaseous/liquid system (e.g., diffusivity, viscosity, density) and the absorber operating conditions (e.g., temperature, flow rates of the gaseous and liquid streams). It is enhanced by lower temperatures, greater contacting surface, higher liquid-gas ratios, and higher concentrations in the gas stream.

Absorption can be physical or chemical. Physical absorption occurs when the absorbed compound simply dissolves in the solvent. When there is a reaction between the absorbed compound and the solvent, it is termed chemical absorption. Liquids commonly used as solvents for organic and inorganic compounds include water, mineral oils, nonvolatile hydrocarbon oils, and aqueous solutions (e.g., sodium hydroxide).

The types of equipment commonly used for gas/liquid contact operations include packed towers, plate or tray towers, spray chambers, and venturi scrubbers. These devices are designed to provide maximum contact between the gas and liquid streams in order to increase the mass transfer rate between the two phases. A packed tower is filled with packing material that is designed to expose a large wetted surface area to the gas stream. Plate towers use plates or trays that are arranged so that the gas stream is dispersed through a layer of liquid on each plate. Bubble-cap plates have been widely used; other types of plates include perforated trays and valve trays. In a spray tower, the gas mixture is contacted with a liquid spray. In a venturi scrubber, the gas and liquid streams come into contact at the throat of the venturi nozzle; venturi scrubbers are typically used for removal of particulate matter (see Section 4.11).

Several different configurations of absorber systems are used for controlling vapor emissions. The simplest configuration is one in which the solvent (usually water) is used on a once-through basis, and is then either discharged to a wastewater treatment system or introduced as a process water stream (see Figure 4-15). The possibility of using solvents other than water on a once-through basis may exist when fresh solvent is available in large quantities as a process raw material or fuel. Another configuration involves using the solvent (usually water) on a once-through basis and stripping it (reverse of absorption) before discharging. In yet another configuration, an organic liquid is used as a solvent and recycled to the absorber after being stripped.

The efficiency of absorption for removing pollutants from gaseous streams depends on several factors, including (a) solubility of the pollutant in a given solvent, (b) concentration, (c) temperature, (d) flow rates of gaseous and liquid streams (liquid to gas ratio), (e) contact surface area, and (f) efficiency of stripping (if solvent is recycled to the absorber).

Figure 4-15. A typical countercurrent packed column absorber system.

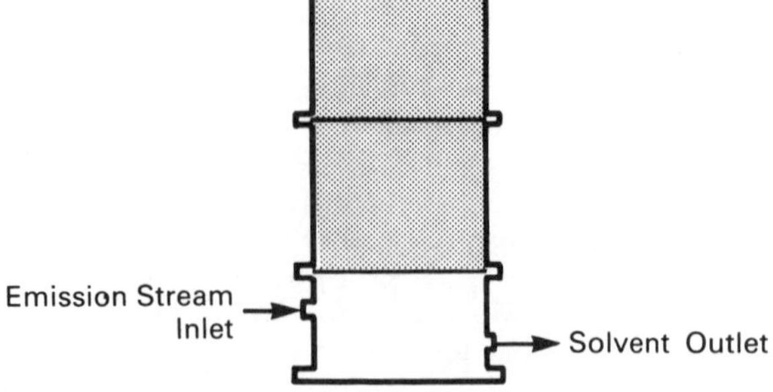

Determination of the absorber system variables (absorber column diameter, height, etc.) is dependent on the individual vapor/liquid equilibrium relationship for the specific HAP/solvent system and the type of absorber to be used (packed or plate tower, etc.). Note that equilibrium data may not be readily available for uncommon HAP's.

Detailed design procedures for all types of absorbers are not appropriate for this handbook; therefore, important design considerations for one type of absorber will be briefly discussed. Since packed towers are commonly used in air pollution control, the discussion will be based on packed tower absorbers. For illustration purposes, a simple configuration is chosen for the absorber system: a packed tower absorber using 2-inch ceramic Raschig rings as the packing material with water used as the absorbent on a once-through basis. The effluent from the absorber is assumed to be discharged to a wastewater treatment facility. The treatment in the following subsections is equally applicable to both organic and inorganic vapor emissions control. (For more information on gas absorption, see references 18, 20, 21, 22, and 23).

As indicated in Chapter 3, absorption is the most widely used control method for inorganic vapor emissions; therefore, Emission Stream 5 containing inorganic vapors will be used in the example case to illustrate the calculation procedures. Worksheets for calculations are provided in Appendix C.7.

4.7.1 Data Required

The data necessary to perform the calculations consist of HAP emission stream characteristics previously compiled on the HAP Emission Stream Data Form and the required HAP control as determined by the applicable regulations.

Example Case

Maximum flow rate, Q_e = 3,000 scfm
Temperature, T_e = 85°F
HAP = ammonia
HAP concentration, HAP_e = 20,000 ppmv
Pressure, P_e = 760 mm Hg

Based on the control requirements for the emission stream:
Required removal efficiency, RE = 98%

In the case of a permit review for an absorber, the following data outlined below should be supplied by the applicant. The calculations in this section will then be used to check the applicant's values.

Absorption system variables at standard conditions (70°F, 1 atm):

Reported removal efficiency, $RE_{reported}$, %
Emission stream flow rate, Q_e, scfm
Temperature of emission stream, T_e, °F
HAP
HAP concentration, HAP_e, ppmv
Solvent used
Slope of the equilibrium curve, m
Solvent flow rate, L_{gal}, gal/min
Density of the emission stream, ρ_G, lb/ft^3
Schmidt No. for the HAP/emission stream and HAP/solvent systems, Sc_G, Sc_L (To calculate Sc_G or Sc_L, see references 18 or 23 for viscosity, density, and diffusivity data.)
Properties of the solvent:
 Density, ρ_L, lb/ft^3
 Viscosity, μ_L, centipoise
Type of packing used
Packing constants a, b, c, d, ϵ, Y, s, g, r
Column diameter, D_{column}, ft
Tower height (packed), Ht_{column}, ft
Pressure drop, ΔP_{total}, in H_2O

4.7.2. Absorption System Design Variables

In absorption, the removal efficiencies (or outlet concentrations) are limited by the driving force available from gas to the liquid phase. The driving force for a given set of operating conditions is determined by the difference between the actual HAP concentrations in the gas stream and solvent and the corresponding equilibrium concentrations.

When the slope (m) of the equilibrium curve is small for a given HAP/solvent system, indicating that the HAP is readily soluble in the solvent, the driving force for absorption is large and absorption occurs readily. On the other hand, if m is large (e.g., > 50), the HAP is not readily soluble in the solvent and the driving force for absorption is small; therefore, long contact times, tall absorption towers, and/or high liquid-gas ratios are required for adequate performance (high removal efficiency and/or low outlet concentrations). Hence, as a conservative guideline, assume that if m is greater than about 50 for a given HAP/solvent system at atmospheric pressure, then high removal efficiencies (~99 percent) are not possible.

4.7.3 Determination of Absorber System Design and Operating Variables

In most applications involving the absorption of a gaseous pollutant from an effluent gas stream, the inlet conditions (flow rate, composition, and temperature) are usually known. The composition of the outlet gas is specified by the control requirements. The conditions of the inlet liquid are also known. The main objectives, then, in the design of an absorption column will be the determination of the solvent flow rate and the calculation of the principal dimensions of the equipment (column diameter and height to accomplish the absorption operation) for a selected solvent.

To keep the discussion simple, the following assumptions are made: (1) there are no heat effects associated with the absorption operation, and (2) both the gas and liquid streams are dilute solutions (i.e., flow rates are constant throughout the absorption column and the equilibrium curve can be approximated as a straight line). All of the data (e.g., packing factors, Schmidt numbers, etc.) required in the calculation of the design variables can be found in references 12, 18, 21, 23, 24, or Appendix B.9, reference 8.

4.7.3.1 Solvent Flow Rate

The quantity of solvent to be used is typically estimated from the minimum liquid-gas ratio as determined from material balances and equilibrium considerations. As a rule of thumb for purposes of rapid estimates, it has frequently been found that the most economical value for the absorption factor (defined below) will be in the range from 1.25 to 2.0.(20)

$$AF = L_{mol}/m\,G_{mol} \qquad (4.7\text{-}1)$$

where:
AF = absorption factor
L_{mol} = liquid (solvent) flow rate, lb-moles/hr
G_{mol} = gas stream flow rate, lb-moles/hr
m = slope of the equilibrium curve

The value of m is determined from the equilibrium data at a specific temperature level for the HAP/solvent system under consideration. (See references 12, 21 and 24 for equilibrium for specific systems. For information on other systems, see references 18, 22, and 23). Assuming a value of 1.6 for AF, use Equation 4.7-1 to calculate the solvent flow rate:

$$L_{mol} = 1.6\,m\,G_{mol} \qquad (4.7\text{-}2)$$

The variable G_{mol} can be expressed in terms of Q_e as follows:

$$G_{mol} = 0.155\,Q_e \qquad (4.7\text{-}3)$$

Note that L_{mol} can be converted to gal/hr basis as follows:

$$L_{gal} = [L_{mol} \times MW_{solvent} \times (1/\rho_L) \times 7.48]/60 \qquad (4.7\text{-}4)$$

where:
L_{gal} = solvent flow rate, gal/min
$MW_{solvent}$ = molecular weight of solvent, lb/lb-mole
ρ_L = density of solvent (liquid), lb/ft^3

The factor 7.48 is used to convert from ft^3 to gal basis. For water as the solvent, ρ_L = 62.43 lb/ft^3 (reference 18) and $MW_{solvent}$ = 18 lb/lb mole; then:

$$L_{gal} = 0.036\,L_{mol} \qquad (4.7\text{-}5)$$

> **Example Case**
> *Using Equations 4.7-2, -3, and -5:*
> $m = 1.3$
> (for the operating conditions in the system, consult reference 21)
> $Q_e = 3{,}000$ scfm
>
> $G_{mol} = 0.155 \times 3{,}000 \times 60$
> $G_{mol} = 465$ lb-moles/hr
>
> $L_{mol} = 1.6 \times 1.3 \times 465$
> $L_{mol} = 970$ lb-moles/hr
>
> $L_{gal} = 0.036 \times 970$
> $L_{gal} = 35$ gal/min

4.7.3.2 Column Diameter

Once the gas and liquid streams entering and leaving the absorber column and their concentrations are identified, flow rates calculated, and operating conditions (type of packing) determined, the physical dimensions of the column can be calculated. The column must be of sufficient diameter to accommodate the gas and liquid streams.

The calculation of the column diameter is based on flooding considerations, the usual operating range being taken as 60 to 75 percent of the flooding rate. One of the commonly used correlations in determining the column diameter is shown in Figure 4-16.(12) The procedure to calculate the column diameter is as follows: First, calculate the abscissa (ABS):

$$ABS = (L/G)\,(\rho_G/\rho_L)^{0.5} \qquad (4.7\text{-}6)$$

where:
L = solvent flow rate, lb/hr
G = gas stream flow rate, lb/hr
ρ_G = density of emission stream, lb/ft^3

The values for the variables L and G can be calculated by multiplying L_{mol} and G_{mol} with their respective molecular weights. Then proceed to the flooding line in Figure 4-16 and read the ordinate (ORD), and solve the ordinate expression for $G_{area,f}$ at flooding:

$$ORD = [(G_{area,f})^2\,(a/\epsilon^3)\,(\mu_L{}^{0.2})]\,/\rho_G\rho_L g_c \qquad (4.7\text{-}7)$$

Thus,

$$G_{area,f} = \{[ORD\ \rho_G\ \rho_L g_c]/[(a/\epsilon^3)\,(\mu_L{}^{0.2})]\}^{0.5} \qquad (4.7\text{-}8)$$

where:
$G_{area,f}$ = gas stream flow rate based on column cross sectional area (at flooding conditions), lb/ft^2-sec
a,ϵ = packing factors (see reference 21)
μ_L = viscosity of solvent, centipoises
g_c = gravitational constant, ft/sec^2

Assuming f as the fraction of flooding velocity appropriate for the proposed operation, the gas

Figure 4-16. Correlation for flooding rate in randomly packed towers.(12)

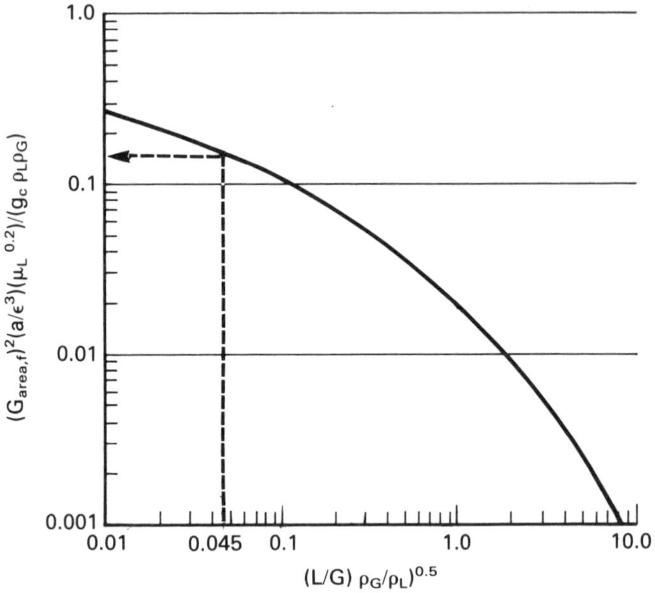

> **Example Case**
> *Using Equations 4.7-6, -7, -8, -9, -10, and -11:*
> $L = MW_{solvent} \times L_{mol} = 18 \times 970$
> $= 17{,}460$ lb/hr
> $G = MW_e \times G_{mol} = 28.4 \times 465 = 13{,}200$ lb/hr
> (see Appendix B.1, reference 8, for calculating MW_e)
> $\rho_G = 0.071$ lb/ft^3 (from ideal gas law at 85°F)
> (see reference 21 for calculating ρ_G)
> $\rho_L = 62.18$ lb/ft^3 (reference 18; at 85°F)
> $ABS = (17{,}460/13{,}200)\,(0.071/62.18)^{0.5} = 0.045$
>
> From Figure 4-16, at ABS = 0.045, the value of ORD at flooding conditions is about 0.15. For 2-inch ceramic Raschig rings, from reference 21:
> $a = 28$
> $\epsilon = 0.74$
>
> Also,
> $g_c = 32.2$ ft/sec^2
> $\mu_L = 0.85$ cp (reference 18, at 85°F)
>
> Thus,
> $$G_{area,f} = \left[\frac{(0.15 \times 0.071 \times 62.18 \times 32.2)}{[28/(0.74)^3]\,(0.85)^{0.2}}\right]^{0.5}$$
> $G_{area,f} = 0.56$ lb/sec-ft^2 (at flooding)
>
> Assuming $f = 0.60$
> $G_{area} = 0.60 \times 0.56 = 0.34$ lb/sec-ft^2
>
> Thus,
> $A_{column} = 13{,}200/(3{,}600 \times 0.34)$
> $A_{column} = 10.8$ ft^2
> $D_{column} = 1.13\,(10.8)^{0.5} = 3.7 \sim 4$ ft

stream flow rate (based on cross-sectional area) can be expressed as:

$$G_{area} = f\, G_{area,f} \qquad (4.7\text{-}9)$$

The usual column operating range for f is taken as 0.60 to 0.75. Calculate the column cross-sectional area by the following expression:

$$A_{column} = g/(3{,}600\, G_{area}) \qquad (4.7\text{-}10)$$

The column diameter is then determined by:

$$D_{column} = (4/\pi)(A_{column})^{0.5} = 1.13(A_{column})^{0.5} \qquad (4.7\text{-}11)$$

where:
D_{column} = column diameter, ft

4.7.3.3 Column Height

The column must be of sufficient height to ensure that the required removal efficiency is achieved. The height of a packed column is calculated by determining the required number of theoretical transfer units and multiplying by the height of a transfer unit. A transfer unit is a measure of the difficulty of the mass transfer operation and is a function of the solubility and concentrations of the solute in the gas and liquid streams. It is expressed as N_{OG} or N_{OL} depending on whether the gas film or liquid film resistance controls the absorption rate. In emission control applications, gas film resistance will typically be controlling, therefore N_{OG} will be used in the following calculations.

The expression for the column height (packed) is:

$$Ht_{column} = N_{OG}\, H_{OG} \qquad (4.7\text{-}12)$$

where:
Ht_{column} = packed column height, ft
N_{OG} = number of gas transfer units (based on overall gas film coefficients)
H_{OG} = height of an overall gas transfer unit (based on overall gas film coefficients), ft

Although the determination of N_{OG} is usually complicated, when dilute solutions are involved, N_{OG} can be calculated using the following equation:

$$N_{OG} = \ln\{(HAP_e/HAP_o)[1 - (1/AF)] + (1/AF)\}/[1 - (1/AF)] \qquad (4.7\text{-}13)$$

This expression is simplified based on the assumption that no HAP is present in the solvent as it enters the column (see reference 8 for details). Alternatively, use Figure 4-17 directly to determine N_{OG}.

The variable H_{OG} is generally calculated from the following equation:

$$H_{OG} = H_G + (1/AF)\, H_L \qquad (4.7\text{-}14)$$

where:
H_G = height of a gas transfer unit, ft
H_L = height of a liquid transfer unit, ft

Figure 4-17. N_{OG} for absorption columns with constant absorption factor AF.(12)

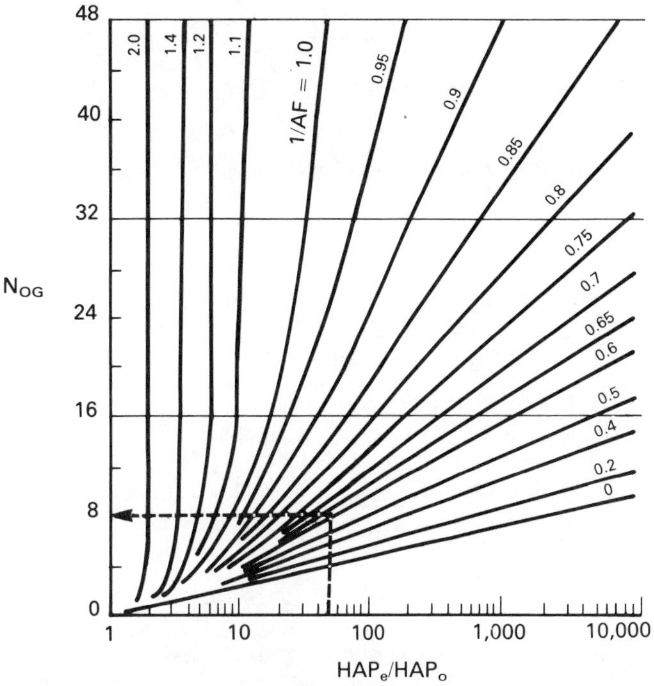

Generalized correlations are available to calculate H_G and H_L; these are based on the type of packing and the gas and solvent flow rates. The correlations for H_G and H_L are as follows:(12)

$$H_G = [b\,(3{,}600\, G_{area})^c/(L'')^d]\,(Sc_G)^{0.5} \qquad (4.7\text{-}15)$$

$$H_L = Y(L''/\mu_L'')^s\,(Sc_L)^{0.5} \qquad (4.7\text{-}16)$$

where:
b, c, d, Y, and s = empirical packing constants (see reference 18)
L'' = liquid flow rate, lb/hr-ft^2
μ_L'' = liquid viscosity, lb/ft-hr
Sc_G = Schmidt number for the gas stream
Sc_L = Schmidt number for the liquid stream

The values for Sc_G and Sc_L are listed for several compounds in references 12 and 23 (or see Appendix B.9, reference 8). In the calculations, it is assumed that the effect of temperature on Sc is negligible. The value for the variable L″ in this equation is calculated as follows:

$$L'' = L/A_{column} \qquad (4.7\text{-}17)$$

Use the following expression to calculate the total column height (Ht_{total}):(24)

$$Ht_{total} = Ht_{column} + 2 + 0.25\, D_{column} \qquad (4.7\text{-}18)$$

Example Case

1. Calculation of N_{OG}

   Using Equation 4.7-13:

   HAP_e = 20,000 ppmv
   HAP_o = 20,000 (1 - 0.98) = 400 ppmv
   N_{OG} = ln[(20,000/400) 0.375 + 0.625]/0.375
   N_{OG} = 7.9

   Using Figure 4 -17:

   HAP_e/HAP_o = 20,000/400 = 50
   At AF = 1.6, 1/AF = 0.63, and N_{OG} = 8

2. Calculation of H_{OG}:

   Using Equations 4.7-14, -15, -16, and -17:

   L″ = 17,460/10.8 = 1,617 lb/hr-ft²
   3,600 G_{area} = 1,224 lb/hr-ft²

   From reference 12, the packing factors are:
   b = 3.82
   c = 0.41
   d = 0.45
   Y = 0.0125
   s = 0.22

   Although 1,224 lb/hr-ft² is outside the range shown in the table, assume that the packing factors are applicable and the error introduced into the calculations will be negligible. From reference 12:
   Sc_G = 0.66
   Sc_L = 570

   Also,
   μ_L'' = 0.85 x 2.42 = 2.06 lb/ft-hr
   (The factor 2.42 is used to convert from centipoise to lb/ft-hr.)

   Hence,
   $H_G = [3.82\,(1{,}224)^{0.41}/(1{,}617)^{0.45}](0.66)^{0.5}$
   = 2.06
   $H_L = 0.0125\,(1{,}617/2.06)^{0.22}(570)^{0.5}$
   = 1.29
   Using AF = 1.6,
   H_{OG} = 2.06 + (1/1.6) 1.29 = 2.87 ~2.9

3. Calculation of Ht_{column}:

   Using Equation 4.7-12:

   Ht_{column} = 7.9 x 2.9 = 22.9 ~23 ft

4. Calculation of Ht_{total}:

   Using Equation 4.7-18:

   Ht_{total} = 23 + 2 + (0.25 x 4) = 26 ft

For costing purposes, it is necessary to calculate the column weight. Use the following equation:(24)

$$Wt_{column} = (48\,D_{column} \times Ht_{total}) + 39\,(D_{column})^2 \quad (4.7\text{-}19)$$

where:
Wt_{column} = column weight, lb

Also, to determine packing costs, volume occupied by the packing material ($V_{packing}$) has to be calculated. Use the following expression:

$$V_{packing} = (\pi/4)(D_{column})^2 \times Ht_{column} \quad (4.7\text{-}20)$$
$$V_{column} = 0.785(D_{column})^2 \times Ht_{column}$$

Example Case

Using Equation 4.7-19:

D_{column} = 4 ft
Ht_{total} = 26 ft
Wt_{column} = (48 x 4 x 26) + 39(4)²
Wt_{column} = 5,600 lb

Using Equation 4.7-20:

Ht_{column} = 23 ft
$V_{packing}$ = 0.785 x (4)² x 23
$V_{packing}$ = 290 ft³

4.7.3.4 Pressure Drop Through the Column

The pressure drop through a packed column for any combination of liquid and gas flows in the operable range is an important economic consideration in the design of such columns. For a particular packing, the most accurate data will be those available from the manufacturer. For purposes of estimation, use the following correlation:(12)

$$\Delta P_a = (g \times 10^{-8})\,[10^{(r_L/\rho_L)}]\,(3{,}600\,G_{area})^2/\rho_G \quad (4.7\text{-}21)$$

where:
ΔP_a = pressure drop, lb/ft²-ft
g, r = packing constants (see reference 12)

The total pressure drop through the column is then expressed as:

$$\Delta P_{total} = \Delta P_a \times Ht_{column} \quad (4.7\text{-}22)$$

Example Case

Using Equation 4.7-21:

From reference 12:
g = 11.13
r = 0.00295

Also,
L″ = 1,617 lb/hr-ft²
3,600 G_{area} = 3,600 x 0.34 = 1,224 lb/hr-ft²
ρ_G = 0.071 lb/ft³
ρ_L = 62.18 lb/ft³

Thus,

$$\Delta P_a = \frac{[11.13 \times 10^{-8} \times 10^{(0.00295 \times 1{,}617/62.18)}(1{,}224)^2]}{(0.071)}$$

ΔP_a = 2.8 lb/ft²-ft

Using Equation 4.7-22:

Ht_{column} = 23 ft
ΔP_{total} = 2.8 x 23 = 64.4 lb/ft²
ΔP_{total} = 64.4/5.2 = 12 in H_2O
(The factor 5.2 is used to convert from lb/ft² to in H_2O.)

4.7.4 Evaluation of Permit Application

Compare the results from the calculations and the values supplied by the permit applicant using Table 4-9. The calculated values in the table are based on the Example Case. If the calculated values of L_{gal}, D_{column}, Ht_{column}, Ht_{total}, ΔP_{total} Wt_{column}, and $V_{packing}$ are different from the reported values for these variables, the differences may be due to the assumptions involved in the calculations. Therefore, the reviewer may wish to discuss the details of the proposed design with the permit applicant.

If the calculated values agree with the reported values, then the design and operation of the proposed scrubber system may be considered appropriate based on the assumptions made in this handbook.

Table 4-9. Comparison of Calculated Values and Values Supplied by the Permit Applicant for Absorption

	Calculated Value (Example Case)[a]	Reported Value
Solvent flow rate, L_{gal}	35 gal/min	...
Column diameter, D_{column}	4 ft	...
Column height, Ht_{column}	23 ft	...
Total column height, Ht_{total}	26 ft	...
Packing volume, $V_{packing}$	290 ft^3	...
Pressure drop, ΔP_{total}	12 in H_2O	...
Column weight, Wt_{column}	5,600 lb	...

[a] Based on Emission Stream 5.

4.8 Condensation

Condensation is a separation technique in which one or more volatile components of a vapor mixture are separated from the remaining vapors through saturation followed by a phase change (see Figure 4-18). The phase change from gas to liquid can be accomplished in two ways: (a) the system pressure may be increased at a given temperature, or (b) the system temperature may be reduced at constant pressure.

The design and operation of a condenser is significantly affected by the number and nature of the components present in the emission stream. In a two-component vapor system where one of the components is noncondensible (e.g., air), condensation occurs at dew point (saturation) when the partial pressure of the condensible compound (e.g., benzene) is equal to its vapor pressure. In most HAP control applications, the emission stream will contain large quantities of noncondensible and small quantities of condensible compounds. To separate the condensible component from the gas stream at a fixed pressure, the temperature of the gas stream must be reduced. The more volatile a compound (i.e., the lower the normal boiling point), the larger the amount that can

Figure 4-18. Flow diagram for a typical condensation system with refrigeration.

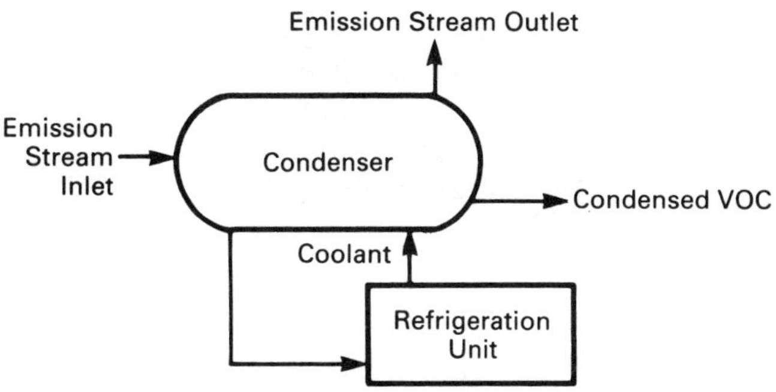

remain as vapor at a given temperature; hence the lower the temperature required for saturation (condensation).

When condensers are used to control emissions, they are usually operated at the constant pressure of the emission source, which is normally close to atmospheric. Depending on the temperatures required for condensation, a refrigeration unit may be necessary to supply the coolant (see Section 4.8.3.2). The two most common types of condensers used are surface and contact condensers. Surface condensers are usually shell-and-tube heat exchangers. The coolant typically flows through the tubes and the vapors condense on the shell outside the tubes. The condensed vapor forms a film on the cool tubes and is drained to a collection tank for storage or disposal. In contrast to surface condensers where the coolant does not contact either the vapors or the condensate, in contact condensers, the vapor mixture is cooled by spraying a cool liquid directly into the gas stream.

Design calculations for condenser systems vary in complexity depending on the nature and number of components present in the emission stream. For detailed information on condenser design, consult references 25 and 26. In the following discussion, Emission Stream 6, consisting of a single condensible component and a single noncondensible component, will be used to illustrate the calculation procedure for surface condensers. It will be assumed that the moisture content of the emission stream is negligible (i.e., no ice is expected to form on the tubes in the condenser). The design procedure will involve determining the condensation temperature required, selection of coolant, and calculation of condenser size and coolant requirements.

4.8.1. Data Required

The data necessary to perform the calculations consist of HAP emission stream characteristics previously compiled on the HAP Emission Stream Data Form and the required HAP control as determined by the applicable regulations.

> **Example Case**
> Maximum flow rate, Q_e = 2,000 scfm
> Temperature, T_e = 90°F
> HAP = styrene
> HAP concentration, HAP_e = 13,000 ppmv (corresponding to saturation conditions)
> Moisture content, M_e = negligible
> Pressure, P_e = 760 mm Hg
>
> Based on the control requirements for the emission stream:
> Required removal efficiency, RE = 90%

In the case of a permit review for a condenser, the following data should be supplied by the applicant. The calculations in this section will then be used to check the applicant's values. Worksheets for calculations are provided in Appendix C.8.

Condenser system variables at standard conditions (70°F, 1 atm):

Reported removal efficiency, $RE_{reported}$, %
Emission stream flow rate, Q_e, scfm
Temperature of emission stream, T_e, °F
HAP
HAP concentration, HAP_e, ppmv
Moisture content, M_e, %
Temperature of condensation, T_{con}, °F
Coolant used
Inlet temperature of coolant, $T_{cool,i}$, °F
Coolant flow rate, $Q_{coolant}$, lb/hr
Refrigeration capacity, Ref, tons
Condenser surface area, A_{con}, ft^2

4.8.2 Pretreatment of the Emission Stream

If water vapor is present in the emission stream, ice may form on the condenser tubes when coolants such as chilled water or brine solutions are used, decreasing the heat transfer efficiency and thus lowering the condenser's removal efficiency. In such cases, dehumidifying the emission stream is necessary. This can be carried out in a heat exchanger prior to the condenser.

> **Example Case**
> Since the moisture content of the emission stream is negligible, no pretreatment is necessary.

4.8.3 Condenser System Design Variables

The key design variable in condenser system design is the required condensation temperature for a given removal efficiency or outlet concentration. A condenser's removal efficiency depends on the nature and concentration of emission stream components. For example, compounds with high boiling points (i.e., low volatility) condense more readily compared to those with low boiling points. Assume, as a conservative starting point, that condensation will be considered as a HAP emission control technique for VOC's with boiling points above 100°F.

The temperature necessary to achieve a given removal efficiency (or outlet concentration) depends on the vapor pressure of the HAP in question at the vapor/liquid equilibrium. Once the removal efficiency for a given HAP is specified, the required temperature for condensation can be determined from data on its vapor pressure-temperature relationship. Vapor pressure-temperature data can be represented graphically (Cox charts) as shown in Figure 4-19 for typical VOC's. The coolant selection is then based on the condensation temperature required. See Table 4-10 for a summary of practical limits for coolant selection.

In a permit evaluation, use Table 4-10 to determine if the values reported for the condensation temperature (T_{con}) and the type of coolant selected are consistent. Also, check if the coolant inlet temperature is based on a reasonable approach temperature (a conservative value of 15°F is used in the

Figure 4-19. Vapor pressure-temperature relationship.

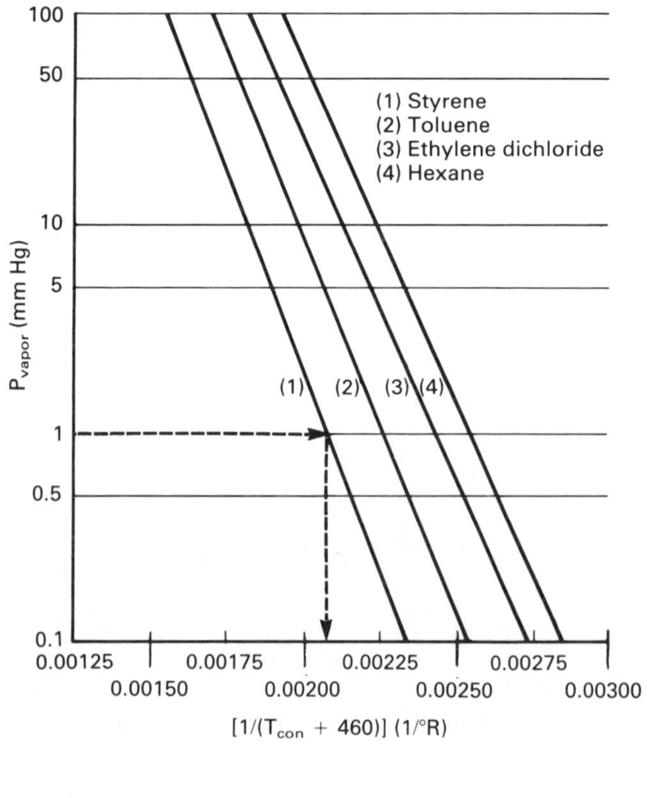

Table 4-10. Coolant Selection

Required Condensation Temperature T_{con} (°F)[a]	Coolant	Coolant Temperature $T_{cool,i}$ (°F)[b]
T_{con}:[c] 60-80	Water	T_{con}-15
$60 > T_{con} > 45$	Chilled water	T_{con}-15
$45 > T_{con} \geq -30$	Brine solutions (e.g., calcium chloride, ethylene glycol)	T_{con}-15
$\sim -90\ T_{con} < -30$	Chlorofluorocarbons (e.g., Freon-12)	T_{con}-15

[a] Also emission stream outlet temperature.
[b] Assume the approach as 15°F.
[c] Summer limit.

table). If the reported values are appropriate, proceed with the calculations. Otherwise, the applicant's design is considered unacceptable. The reviewer may then follow the calculation procedure outlined below.

4.8.4 Determining Condenser System Design Variables

The condenser system evaluated in this handbook consists of a shell-and-tube heat exchanger with the hot fluid (emission stream) in the shell side and the cold fluid (coolant) in the tube side. The emission stream is assumed to consist of a two-component mixture: one condensible component (HAP) and one noncondensible component (air). Typically, condensation for such a system occurs nonisothermally. To simplify the calculations, it is assumed that condensation occurs isothermally.

4.8.4.1 Estimating Condensation Temperature

In the following calculations, it is assumed that the emission stream entering the condenser consists of air saturated with the HAP in question. Calculations for cases involving mixtures of HAP's and supersaturated streams are quite complex and will not be treated here; for additional information, consult references 25 and 26.

For a given removal efficiency, the first step in the calculation procedure is to determine the concentration at the outlet of the condenser. Use the following expression:

$$P_{partial} = 760\,\{(1 - 0.01\ RE)/[(1 - (RE \times 10^{-8} \times HAP_e)]\}\ HAP_e \times 10^{-6} \qquad (4.8\text{-}1)$$

where $P_{partial}$ is the partial pressure (mm Hg) of the HAP in the exit stream assuming the pressure in the condenser is constant and at atmospheric. At equilibrium between the gas and liquid phases, the partial pressure of the HAP is equal to its vapor pressure at that temperature. Therefore, by determining the temperature at which this condition occurs, the condensation temperature (T_{con}) can be specified. To carry out this calculation, vapor pressure-temperature data for the specific HAP are required (see Figure 4-19). Such data can be obtained from references 18 and 27.

Example Case

Using Equation 4.8-1 and Figure 4-19:

HAP_e = 13,000 ppmv (styrene)
RE = 90%
$P_{partial}$ = $760\,\{[1-(0.01 \times 90)]/[(1-(90 \times 10^{-8} \times 13{,}000)]\}\ 13{,}000 \times 10^{-6}$
$P_{partial}$ = 1.0 mm Hg

For styrene, the value of $[1/(T_{con} + 460)]$ corresponding to 1.0 mm Hg in Figure 4-19 is about 0.00208.

Solving for T_{con};
T_{con} = 20°F

4.8.4.2 Selecting the Coolant

The next step is to select the coolant based on the condensation temperature required. Use Table 4-10 to specify the coolant type. For additional information on coolants and their properties, see references 18 and 27.

Example Case

Based on T_{con} = 20°F, the appropriate coolant is a brine solution. Assume the brine solution is a 29% (wt) calcium chloride solution which can be cooled down to -40°F (see reference 18).

4.8.4.3 Condenser Heat Load

Condenser heat load is defined as the quantity of heat that must be extracted from the emission stream to achieve a certain level of removal. It is determined from an energy balance, taking into account the heat of condensation of the HAP, sensible heat change of the HAP, and the sensible heat change in the emission stream. The calculation steps are outlined below:

1. a. Calculate moles of HAP in the inlet emission stream (Basis: 1 min):

$$HAP_{e,m} = (Q_e/387)\ HAP_e \times 10^{-6} \qquad (4.8\text{-}2)$$

The factor 387 is the volume (ft^3) occupied by 1 lb-mole of ideal gas at standard conditions (70°F and 1 atm).

b. Calculate moles of HAP remaining in the outlet emission stream (Basis: 1 min):

$$HAP_{o,m} = (Q_e/387)[1 - (HAP_e \times 10^{-6})][P_{vapor}/(P_e - P_{vapor})] \quad (4.8\text{-}3)$$

where P_{vapor} is equal to $P_{partial}$.

c. Calculate moles of HAP condensed (Basis: 1 min):

$$HAP_{con} = HAP_{e,m} - HAP_{o,m} \quad (4.8\text{-}4)$$

2. a. Determine the HAP's heat of vaporization (ΔH): Typically the heat of vaporization will vary with temperature. Using vapor pressure-temperature data as shown in Figure 4-19, ΔH can be estimated by linear regression for the vapor pressure and temperature range of interest. (See references 18 and 19 for details.)

b. Calculate the enthalpy change associated with the condensed HAP (Basis: 1 min):

$$H_{con} = HAP_{con} [\Delta H + \bar{C}p_{HAP} (T_e - T_{con})] \quad (4.8\text{-}5)$$

where $\bar{C}p_{HAP}$ is the average specific heat of the HAP for the temperature interval T_{con} - T_e (Btu/lb-mole-°F).

c. Calculate the enthalpy change associated with the uncondensed HAP (Basis: 1 min):

$$H_{uncon} = HAP_{o,m} \bar{C}p_{HAP} (T_e - T_{con}) \quad (4.8\text{-}6)$$

d. Calculate the enthalpy change associated with the noncondensible vapors (i.e., air) (Basis: 1 min):

$$H_{noncon} = [(Q_e/387) - HAP_{e,m}] \bar{C}p_{air} (T_e - T_{con}) \quad (4.8\text{-}7)$$

where $\bar{C}p_{air}$ is the average specific heat of air for the temperature interval T_{con} - T_e (Btu/lb-mole-°F).

3. a. Calculate the condenser heat load (Btu/hr) by combining Equations 4.8-5, -6, and -7:

$$H_{load} = 1.1 \times 60 (H_{con} + H_{uncon} + H_{noncon}) \quad (4.8\text{-}8)$$

The factor 1.1 is included as a safety factor.

4.8.4.4. Condenser Size

Condenser systems are typically sized based on the total heat load and the overall heat transfer coefficient estimated from individual heat transfer + coefficients of the gas stream and the coolant. An accurate estimate of individual coefficients can be made using physical/chemical property data for the gas stream, the coolant, and the specific shell-and-tube system to be used. Since this level of detail is not appropriate here, the value used for the overall heat transfer coefficient is a conservative estimate.

Example Case

Using Equations 4.8-2 to -8:

1. a. Q_e = 2,000 scfm
   $HAP_{e,m}$ = (2,000/387) 13,000 x 10^{-6}
   $HAP_{e,m}$ = 0.06718 lb-moles/min

   b. P_{vapor} = 1.0 mm Hg
   P_e = 760 mm Hg
   $HAP_{o,m}$ = 5.1008 [1.0/(760 - 1.0)]
   $HAP_{o,m}$ = 0.00672 lb-moles/min

   c. HAP_{con} = .06718 - 0.00672
   = 0.0605 lb-moles/min

2. a. ΔH = 17,445 Btu/lb-mole
   (see Appendix B.10, reference 8, or references 18 and 19)

   b. MW_{HAP} = 104.2 lb/lb-mole
   $\bar{C}p_{HAP}$ = 24 Btu/lb-mole-°F
   (extrapolated from data in reference 27)
   H_{con} = 0.0605 [17,445 + 24 (90-20)]
   H_{con} = 1,157 Btu/min

   c. H_{uncon} = 0.00672 x 24 x (90 - 20)
   H_{uncon} = 11.3 Btu/min

   d. $\bar{C}p_{air}$ = 6.96 Btu/lb-mole-°F
   (see reference 3 for details)
   H_{noncon} = [(2,000/387) - 0.06718] 6.96 x (90-20)
   H_{noncon} = 2,485 Btu/min

3. a. H_{load} = 1.1 x 60 (1,157 + 11.3 + 2,485)
   H_{load} = 241,100 Btu/hr

For additional information on how to calculate individual heat transfer coefficients, consult reference 25.

To size condensers, use the following equation to determine the required heat transfer area:

$$A_{con} = H_{load}/U\Delta T_{LM} \quad (4.8\text{-}9)$$

where:

A_{con} = condenser (heat exchanger) surface area, ft^2
U = overall heat transfer coefficient, Btu/hr-ft^2-°F
ΔT_{LM} = logarithmic mean temperature difference, °F

and:

$$\Delta T_{LM} = \frac{(T_e - T_{cool,o}) - (T_{con} - T_{cool,i})}{\ln [(T_e - T_{cool,o}) - (T_{con} - T_{cool,i})]}$$

where:

T_e = emission stream temperature, °F
$T_{cool,o}$ = coolant outlet temperature, °F
T_{con} = condensation temperature, °F
$T_{cool,i}$ = coolant inlet temperature, °F

Assume that the approach temperature at the condenser exit is 15°F. In other words, $T_{cool,i} = (T_{con} - 15)$. Also, the temperature rise of the coolant fluid is specified as 25°F, i.e., $T_{cool,o} = (T_{cool,o} + 25)$ where $T_{cool,o}$ is the coolant exit temperature. In estimating A_{con}, the overall heat transfer coefficient can be conservatively assumed as 20 Btu/hr-ft^2-°F; the actual value will depend on the specific system under consideration. This is based on reference 26 in which guidelines on typical overall heat transfer coefficients for condensing vapor-liquid media are reported.

Example Case

Using Equation 4.8-9:

T_e = 90°F
T_{con} = 20°F
$T_{cool,i}$ = 20 - 15 = 5°F
$T_{cool,o}$ = $T_{cool,i}$ + 25 = 30°F

$$\Delta T_{LM} = \left[\frac{(90 - 30) - (20 - 5)}{\ln [(90 - 30)/(20 - 5)]} \right]$$

ΔT_{LM} = 32°F

H_{load} = 241,100 Btu/hr
U = 20 Btu/hr-ft^2-°F

A_{con} = 241,100/(20 x 32)
A_{con} = 375 ft^2

4.8.4.5 Coolant Flow Rate

The quantity of heat extracted from the emission stream is transferred to the coolant. By a simple energy balance, the flow rate of the coolant can be calculated as follows:

$$Q_{coolant} = H_{load}/[\underline{C}p_{coolant} (T_{cool,o} - T_{cool,i})] \qquad (4.8\text{-}10)$$

where:

$Q_{coolant}$ = coolant flow rate, lb/hr
$\underline{C}p_{coolant}$ = average specific heat of the coolant over the temperature interval $T_{cool,i}$ to $T_{cool,o}$, Btu/lb-°F

Specific heat data for coolants are available in references 18 and 27.

Example Case

Using Equation 4.8-10:

H_{load} = 241,100 Btu/hr
$T_{cool,i}$ = 5°F
$T_{cool,o}$ = 30°F
$\underline{C}p_{coolant}$ = 0.65 Btu/lb-°F (reference 18)
$Q_{coolant}$ = 241,100/[0.65 (35-10)]
$Q_{coolant}$ = 14,840 lb/hr

4.8.4.6 Refrigeration Capacity

A refrigeration unit is assumed to supply the coolant at the required temperature to the condenser. For costing purposes, the required refrigeration capacity is expressed in terms of refrigeration tons as follows:

$$Ref = H_{load}/12{,}000 \qquad (4.8\text{-}11)$$

where Ref is the refrigeration capacity, tons.

Example Case

Using Equation 4.8-11:

H_{load} = 241,100 Btu/hr
Ref = 241,100/12,000
Ref = 20 tons

4.8.4.7 Recovered Product

To calculate costs, the quantity of recovered product that can be sold and/or recycled to the process must be determined. Use the following equation:

$$Q_{rec} = 60 \times HAP_{con} \times MW_{HAP} \qquad (4.8\text{-}12)$$

where Q_{rec} is the quantity of product recovered, lb/hr.

Example Case

Using Equation 4.8-12:

HAP_{con} = 0.0605 lb-moles/min
MW_{HAP} = 104.2 lb/lb-mole
Q_{rec} = 60 x 0.06065 x 104.2
Q_{rec} = 378 lb/hr

4.8.5 Evaluation of Permit Application

Compare the results from the calculations and the values supplied by the permit applicant using Table 4-11. The calculated values in the table are based on the Example Case. If the calculated values of T_{con}, coolant type, A_{con}, $Q_{coolant}$, Ref, and Q_{rec} are different from the reported values for these variables, the differences may be due to the assumptions involved in the calculations. Therefore, the reviewer may wish to discuss the details of the proposed design with the permit applicant.

Table 4-11. Comparison of Calculated Values and Values Supplied by the Permit Applicant for Condensation

	Calculated Value (Example Case)[a]	Reported Value
Condensation temperature, T_{con}	20°F	...
Coolant type	Brine solution	...
Coolant flow rate, $Q_{coolant}$	14,840 lb/hr	...
Condenser surface area, A_{con}	375 ft^2	...
Refrigeration capacity, Ref	20 tons	...
Recovered product, Q_{rec}	378 lb/hr	...

[a] Based on Emission Stream 6.

If the calculated values agree with the reported values, then the design and operation of the proposed condenser system may be considered appropriate based on the assumptions made in this handbook.

4.9 Fabric Filters

Fabric filter collectors (also known as baghouses) are one of the most efficient means of separating particulate matter from a gas stream. Fabric filters are capable of maintaining mass collection efficiencies of greater than 99 percent down to a particle size approaching 0.3 μm in most applications.(28,29,30) This efficiency is largely insensitive to the physical characteristics of the gas and dust, and, depending on fabric cleaning method, to the inlet dust loading.(28,31) Physical limitations of the fabric materials to the temperature, moisture content, and corrosivity of the gas stream reduce the applicability of fabric filters. Variables considered in baghouse design include fabric type, cleaning method, air-to-cloth ratio, and equipment configuration. The filter fabric, cleaning method, and air-to-cloth ratio all should be selected concurrently; choice of these parameters is mutually dependent.(28) Equipment configuration is of secondary importance unless site-specific space limitations exist that require configuration to be of primary importance in the fabric filter design.

Fabric filter systems typically are designed on the basis of empirical information obtained through testing and long-term actual operating experience for similar combinations of cleaning method, fabric type, and dust rather than by analytical methods.(28) Although theoretical equations exist to predict the performance of filtering systems under various conditions, these equations are not very useful as design tools. Therefore, discussion of baghouse design in this section provides qualitative guidance rather than predictive equations. Generally, fabric filter design for HAP's is no different than fabric filter design for control of any other type of particulate matter. However, due to the hazards associated with HAP's, greater care must be taken to ensure that control is consistently of high efficiency and that the control device is leak-proof, thus preventing accidental release of the gas stream and captured pollutants. For these reasons, design of a fabric filter for HAP's should only consider selected fabric cleaning methods, and the design should specify an induced draft fan (i.e., a negative pressure or suction baghouse) rather than a forced draft fan (i.e., a positive pressure baghouse). Information presented in this section can be used to provide guidance for or to evaluate the appropriateness of baghouse design for certain HAP applications.

Appendix C.9 provides a worksheet to record the information obtained during the performance of the fabric filter design procedures.

4.9.1 Data Required

The data necessary to perform the design steps consist of the HAP emission stream characteristic previously compiled on the HAP Emission Stream Data Form and the required HAP control as determined by the applicable regulations.

Example Case

Fabric filtration was one of the selected control techniques for the municipal incinerator. The pertinent data for these procedures are found on the HAP Emission Stream Data Form (see Figure 3-8).

Flow rate, $Q_{e,a}$ = 110,000 acfm
Moisture content, Me = 5% vol
Temperature, T_e = 400°F
Particle mean diam. = 1.0 μm
SO_3 content = 200 ppm (vol)
Particulate content = 3.2 grains/scf - flyash
HAP content = 10% (mass) cadmium

In the case of a permit review for a fabric filter, the following data should be supplied by the applicant:
Filter fabric material
Cleaning method
Air-to-cloth ratio, ft/min
Baghouse configuration

The design criteria and considerations discussed in this section will be used to evaluate the reasonableness of the applicant's proposed design.

4.9.2 Pretreatment of the Emission Stream

As discussed in Section 3.3.1, the temperature of the emission stream should be within 50° to 100°F above the stream dew point. Procedures for determining the dew point of an emission stream are provided in Appendix B.1. If the emission stream temperature does not fall within the stated range, pretreatment (i.e., emission stream preheating or cooling) is necessary. Methods of pretreatment are briefly discussed in reference 32. If pretreatment is performed, the emission stream characteristics will be altered. The primary characteristics affecting baghouse design are emission stream temperature and flow rate. Therefore, after selecting a temperature for the emission stream, the new stream flow rate must be calculated. The calculation method depends upon the type of pretreatment performed; use appropriate standard industrial equations. The use of pretreatment mechanical dust collectors may also be appropriate. If the emission stream contains an appreciable amount of large particles

(20 to 30 μm), pretreatment with mechanical dust collectors is typically performed. (Appendix B.11, reference 8, further describes the use of mechanical dust collectors.)

4.9.3 Fabric Filter System Design Variables

Successful design of a fabric filter depends on the proper selection of fabric and cleaning method and on an adequate air-to-cloth ratio. All fabric filter systems share the same basic features and operate using the principle of aerodynamic capture of particles by fibers. Systems vary, however, in certain key details of construction and in the operating parameters. The design variables of particular interest are filter bag material, fabric cleaning method, air-to-cloth ratio, baghouse configuration, and materials of construction.

As stated earlier, the first three variables should be considered concurrently. The configuration and construction materials are important, but secondary, considerations. The following subsections discuss step-by-step procedures for selecting each of these design variables as they may apply to a specific particulate HAP control situation. Because HAP control is similar to particulate control in general, a good verification of these procedures can be accomplished by consulting the section about the particular industry in a document entitled *Control Techniques for Particulate Emissions from Stationary Sources—Volume 2,* or in the McIlvaine Fabric Filter Manual.(31,33) (Note: Because these design variables are considered concurrently, the Example Case is presented at the end of Section 4.9.3.)

4.9.3.1 Fabric Type

Several types of natural and synthetic fabric are used in baghouse systems. Gas stream characteristics such as temperature, acidity, alkalinity, and particulate matter properties (e.g., abrasiveness and hygroscoposity), determine the fabric type to be used.(28,34) In many instances, several fabric types will be appropriate, and a final selection will be chosen only when cleaning method and the desired air-to-cloth ratio are considered.

Most of the principal synthetic fibers have been adapted for use as filtering fabrics while the only natural fibers in common use are cotton and wool. Some of the more common synthetic fibers in commercial use are nylon (aromatic and polyamide), acrylic, polyester, polypropylene, fiberglass, and fluorocarbon. Natural fibers can be used for gas temperatures up to 200°F and have only moderate resistance to acids and alkalis contained in the gas stream.(28,31,34) Synthetics can operate at temperatures up to 550°F and generally have greater chemical resistance.(28,31,34) Therefore, while the initial cost of the synthetic filter fabric is greater than the cost for natural fibers, the increased service life and improved operating characteristics of the synthetics make them a preferred choice in a wide range of industrial situations.

Almost all of the filter fabrics can be constructed in either a woven or a felted manner (cotton and fiberglass can be constructed in a woven manner only). Woven fabrics are made up of yarn in one of a variety of patterns that allow spaces between the fibers, whereas felted fabrics are composed of a thick mat of randomly oriented fibers. When woven fabrics are new, particles penetrate the pores of the fabric fairly easily. As filtering continues, however, more particles are retained on the filter threads and on the particles already collected. As this dust layer or "cake" builds up, particle penetration drops to a very low level. Cleaning of woven fabrics must be performed so that a layer of this dust cake remains on the fabric, enabling particle penetration to remain low.(28,33) Felted fabrics are thick enough that a dust cake does not need to remain on the fabric in order to maintain a good collection efficiency.(28,31) This difference between woven and felted fabrics has important implications for selection of fabric cleaning method, as described in Section 4.9.3.2. Felted fabrics are more expensive than woven fabrics.

Table 4-12 presents information on the maximum continuous operating temperature and resistance characteristics of commonly used filter fabrics. Knowing the emission stream characteristics, Table 4-12 can be used to select an appropriate fabric filter type (or types). Although the information presented is qualitative, Table 4-12 provides a good basis either for selecting a fabric or for evaluating the appropriateness of a fabric in a permit application.

When a number of fabrics are suitable for an application, the relative cost of the fabrics may be the key decision criterion. In general, fluorocarbon and nylon aromatic bags are the most expensive, followed by wool and fiberglass. The remaining commonly used synthetics are generally less expensive than fiberglass (polypropylene, polyester, acrylic, nylon polyamide, and modacrylic), while cotton is generally the least expensive fabric.(29,30,31,35)

4.9.3.2 Cleaning Method

As dust accumulates on the filtering elements, the pressure drop across the bag compartment increases until cleaning of the bags occurs. A timer can be used to control the cleaning cycle or pressure drop can be monitored so that cleaning occurs when some maximum desirable value is reached. At this point the bags in the compartment are cleaned to remove the collected dust and the cycle is then repeated. The two basic mechanisms used to accomplish bag cleaning are flexing of the fabric to break up and dislodge the dust cake, and re-

Table 4-12. Characteristics of Several Fibers Used in Fabric Filtration[a]

Fiber Type[b]	Max. Continuous Operating Temp. (°F)	Resistance[c]				
		Abrasion	Mineral Acids	Organic Acids	Alkalies	Solvent
Cotton[d]	180	G	P	G	G	E
Wool[e]	200	F/G	F	F	P/F	G
Modacrylic[e] (Dynel)	175	F	VG	VG	G	G
Polypropylene[e]	200	E	E	E	E	G
Nylon Polyamide[e] (Nylon 6 & 66)	220	E[f]	P	F	VG	E
Acrylic[e] (Orlon)	260	G	G	G	F	E
Polyester[e] (Dacron, Creslan)	275	E	G	G	G	E
Nylon Aromatic[e] (Nomex)	450	E	F	G	VG	E
Fluorocarbon[e] (Teflon, TFE)	500	F/G	E[g]	E[g]	E[g]	E[g]
Fiberglass[d]	550	P,G[h]	VG	E	P	E

[a]References 8, 29, 31, 34, 35, and 37. Where data differed, a representative category was chosen.
[b]Represents the major categories of filtration fibers. Names in parentheses indicate some principal trade names.
[c]P = poor resistance, F = fair resistance, G = good resistance, VG = very good resistance, and E = excellent resistance.
[d]Woven fabrics only.
[e]Woven or felted fabrics.
[f]Considered to surpass all other fibers in abrasion resistance.
[g]The most chemically resistant of all these fibers.
[h]After treatment with a lubricant coating.

versed air flow through the fabric to remove the dust.(28) These may be used separately or in conjunction with one another. The three principal methods used to accomplish fabric cleaning are mechanical shaking (manual or automatic), reverse air flow, and pulse-jet cleaning. The first method uses only the fabric flexing mechanism; the latter two methods use a combination of the reverse air flow and fabric flexing mechanisms.

Selection of a cleaning method is based on the type of fabric used, the pollutant collected, and the manufacturer's, vendor's, and industry's experiences. A poor combination of filter fabric and cleaning method can cause premature failure of the fabric, incomplete cleaning, or blinding of the fabric.(28) Blinding of a filter fabric occurs when the fabric pores are blocked and effective cleaning can not occur. Blinding can result because moisture blocks the pores or increases the adhesion of the dust, or because a high velocity gas stream imbeds the particles too deeply in the fabric.(28) The selection of a cleaning method may be based on cost, especially where more than one method is applicable. Table 4-13 contains a comparison of cleaning methods. Cleaning methods are discussed individually below.

With *mechanical shaking*, bags are hung on an oscillating framework that periodically shakes the bags at timed intervals or at a predefined pressure drop level.(28,30,34) The shaker mechanisms produce a violent action on the fabric filter bags and, in general, produce more fabric wear than the other types of cleaning mechanisms.(30) For this reason, mechanical shaking is used in conjunction with heavier and more durable fabric materials, such as most woven fibers.(30,38) Bags with poor or fair abrasion ratings in Table 4-12 (such as fiberglass) should not be chosen for fabric filters cleaned by mechanical shaking unless they are treated with a special coating before use. Although shaking is abrasive to the fabric, it does allow a dust cake to remain on the fabric, thus maintaining a high collection efficiency.

Bags are usually taken off-line for cleaning by mechanical shaking so that no gas flows through the bags being cleaned. Thus, reentrainment of particles is minimized. Because dust dislodgement is not severe (i.e., a light dust cake remains on the fabric), and because cleaning occurs off-line, outlet concentrations are almost constant with varying inlet dust loading and through entire cleaning cycles when using mechanical shaking.(28) Further control efficiency is very high, and, in fact, properly selected woven fabrics cleaned by mechanical shaking can provide much greater particle collection than pulse-jet cleaned felted fabrics in many applications.(31) For these reasons, mechanical shaking is a good method to clean fabric filters controlling emissions containing HAP's.(31)

Reverse air flow cleaning is used to flex or collapse the filter bags by allowing a large volume of low pressure air to pass countercurrent to the direction

Table 4-13. Comparisons of Fabric Filter Bag Cleaning Methods (31)

Parameter	Cleaning Method: Mechanical Shake	Reverse Airflow	Pulse-jet Individual Bags	Pulse-jet Compartmented Bags
Cleaning On- or Off-line	Off-line	Off-line	On-line	Off-line
Cleaning Time	High	High	Low	Low
Cleaning Uniformity	Average	Good	Average	Good
Bag Attrition	Average	Low	Average	Low
Equipment Ruggedness	Average	Good	Good	Good
Fabric Type	Woven	Woven	Felt	Felt
Filter Velocity	Average	Average	High	High
Power Cost	Low	Low to medium	High	Medium
Dust Loading	Average	Average	Very high	High
Maximum Temperature[a]	High	High	Medium	Medium
Collection Efficiency	Good	Good	Lower	Lower

[a]Fabric limited.

of normal gas stream flow during filtration.(30,34) Reverse air is provided either by a separate fan or by a vent in the fan damper, which allows a backwash of air to clean the fabric filters.(30,34) Reverse air flow cleaning usually occurs off-line. Reverse air cleaning allows the use of fragile bags, such as fiberglass, or light-weight bags, and usually results in longer life for the bags.(30) As with mechanical shaking, woven fabrics are used, and because cleaning is less violent than with pulse-jet cleaning and occurs off-line, outlet concentrations are almost constant with varying inlet dust loading and throughout the cleaning cycle. Reverse air flow cleaning is, therefore, a good choice for fabric cleaning in HAP control situations.

In *pulse-jet cleaning*, a high pressure air pulse is introduced into the bag from the top through a compressed air jet.(30,34) This rapidly expands the bag, dislodging the particles. Thus, the fabric is cleaned thoroughly through a vibration effect. The pulse of air cleans so effectively that no dust cake remains on the fabric to contribute to particulate collection. Because such a cake is essential for effective collection on woven fabrics, felted fabrics are generally used in pulse-jet cleaned fabric filters.(28) All of the fabric materials may be used with pulse-jet cleaning except cotton or fiberglass.

Because the cleaning air pulse is of such high pressure (up to 100 psi) and short duration (≤0.1 sec), cleaning is usually accomplished on-line. Extra bags are not necessary, therefore, to compensate for bags off-line during cleaning. Cleaning occurs more frequently than with mechanical shaking or reverse air flow cleaning, which permits higher air velocities (higher A/C ratios) than the other cleaning methods. Further, because the bags move less during cleaning, they may be packed more closely together. In combination, these features allow pulse-jet cleaned fabric filters to be installed in a smaller space, and thus, at a lower cost, than fabric filters cleaned by the other methods.(28,34) This cost savings may be somewhat counterbalanced by the greater expense and more frequent replacement required of felted bags, the higher power use that may occur, and the installation of the fabric filter framework that pulse-jet cleaning requires.(28,34)

Pulse-jet cleaning is not, however, recommended for HAP control situations for several reasons. First, because cleaning occurs on-line, the rapid high-pressure pulse generated during cleaning causes increased emissions from the bags. Mass emissions can vary by as much as 100 times over a filtration cycle.(31) Second, although collection efficiencies of pulse-jet cleaned fabric filters are in the 99.9 to 99.99 percent range, filtering efficiency of pulse-jet cleaned filters is inferior to that of mechanically shaken or reverse air cleaned filters that have a good cake buildup.(31) In one study, average outlet concentrations were two to three orders of magnitude higher for pulse-jet cleaned filters than for mechanically shaken filters.(31) Third, emissions from pulse-jet systems are strongly dependent on the inlet concentration; thus, the collection efficiency rather than the effluent concentration tends to be relatively constant for fabric filters using pulse-jet cleaning.(28,31) For these reasons, outlet emission levels are not as constant or as low when using pulse-jet cleaning as when using either mechanical shaking or reverse air flow cleaning. Pulse-jet cleaning is, therefore, not recommended for fabric filters used in HAP control situations or for high inlet loadings involving fine particulate matter.(29)

In cases of permit evaluation where pulse-jet cleaning is believed to be adequate to meet specific regulations in specific applications, several options are available to minimize the disadvantages of pulse-jet cleaning. First, pulse-jet filter bags can be compartmentalized to permit off-line cleaning; additional bags must be installed to allow this.(31) Second, reduced filtration velocity, or pulse intensity, will decrease average outlet concentration.(31) Third, bags should be flexible, lightweight, and inelastic, with uniform pore structure, to obtain maximum particle collection.(31) These changes, in effect, alter the typical pulse-jet baghouse such that it behaves (i.e., cleans the bags) in a manner that is very similar to that of a reverse air baghouse.

4.9.3.3 Air-to-Cloth Ratio

The air-to-cloth (A/C) ratio, or filtration velocity, is a traditional fabric filter design parameter defined as the actual volumetric flow rate (acfm) divided by the total active, or net, fabric area (ft^2). The A/C ratio is an important indicator of the amount of air that can be filtered in a given time when considering the dust to be collected, cleaning method and fabric to be used, and the characteristics of the gas stream to be filtered for an individual situation. Selection of an appropriate range of A/C ratios is not based on any theoretical or empirical relationship, but rather is based on industry and fabric filter vendor experience from actual fabric filter installations. A ratio is usually recommended for a specific dust and a specific cleaning method. For typical design calculations, the A/C ratio must be obtained from the literature or the manufacturer.

Table 4-14 summarizes the ranges of recommended A/C ratios by typical bag cleaning method for many dusts and fumes. These ranges are meant to serve as a guide; A/C ratios may vary from those reported. Fabric filter size and cost will vary with A/C ratio; lower A/C ratios, for example, will require that a larger and more expensive fabric filter be installed. (Note: Pulse-jet cleaning is not recommended for HAP control situations; the A/C ratio for control of streams containing HAP's will, therefore, be fairly low.) In addition to evaluating a particular fabric filter application, the A/C ratio and the emission stream flow rate ($Q_{e,a}$) are used to calculate net cloth area (A_{nc}):

$$\frac{Q_{e,a}}{\text{A/C ratio}} = A_{nc} \qquad (4.9\text{-}1)$$

where:

$Q_{e,a}$ = emission stream flow rate at actual conditions acfm
A/C ratio = air-to-cloth ratio, acfm/ft^2 or ft/min
A_{nc} = net cloth area, ft^2

Net cloth area is the cloth area in active use at any point in time. Gross cloth area (A_{tc}), by comparison, is the total cloth area contained in a fabric filter, including that which is out of service at any point in time for cleaning or maintenance. In this handbook, costing of the fabric filter structure uses net cloth area, while costing of fabric filter bags uses gross cloth area. Table 4-15 presents factors to obtain gross cloth area from net cloth area:

$$A_{nc} \times \text{Factor} = A_{tc} \qquad (4.9\text{-}2)$$

where:

Factor = value from Table 4-15, dimensionless
A_{tc} = gross cloth area, ft^2

Fabric filters with a higher A/C ratio require fewer bags to accomplish cleaning, and, therefore, require less space and may be less expensive. Other costs, such as more expensive (felted) bags, bag framework structure, use of increased pressure drop and corresponding increased power requirements, etc., may counterbalance to some degree the savings of high A/C ratio systems.

4.9.3.4 Baghouse Configuration

The basic configuration of a baghouse varies according to whether the gases are pushed through the system by a fan located on the upstream side (forced draft fan), or pulled through by locating the fan on the downstream side (induced draft fan). A baghouse using forced draft fans is called a positive-pressure baghouse; one using induced draft fans is called a negative-pressure or suction baghouse. Positive-pressure baghouses may be either open to the atmosphere or closed (sealed and pressure-isolated from the atmosphere). Negative-pressure baghouses can only be of the closed type. Only the closed suction design should be selected for a HAP application to prevent accidental release of captured pollutants.(34) The higher the gas stream dew point, the greater the precaution that must be taken to prevent condensation, which can moisten the filter cake, plug the cloth, and promote corrosion of the housing and hoppers. In a suction-type fabric filter, infiltration of ambient air can occur, which can lower the temperature below design levels. Therefore, the structure walls and hoppers of this type of baghouse should be insulated to minimize the possibility of condensation.

4.9.3.5 Materials of Construction

The most common material used in fabric filter construction is carbon steel. In cases where the gas stream contains high concentrations of SO_3 or where liquid-gas contact areas are involved, stainless steel may be required. Stainless steel will increase the cost of the fabric filter significantly when compared to carbon steel.(30) However, by keeping the emission stream temperature above the dew point and by insulating the baghouse, the use of stainless steel should not be necessary.

Table 4-14. Recommended Air-to-Cloth (A/C) Ratios for Various Dusts and Fumes by Cleaning Method (28, 36)

Dust or Fume	A/C Ratios Recommended for Cleaning Method (ft/min)		
	Shaker	Reverse Air	Pulse-Jet
Abrasives	2.0 - 3.0	*	9
Alumina	2.25 - 3.0	*	*
Aluminum	3.0	*	16
Aluminum Oxide	2.0	*	*
Asbestos	2.5 - 4.0	*	9 - 16
Bauxite	2.25 - 3.2	*	8 - 10
Blast Cleaning	3.0 - 3.5	*	*
Carbon	1.2 - 2.5	*	5 - 7
Carbon Black	1.5 - 2.5	1.1 - 1.5	8 - 12
Chrome	1.5 - 2.5	*	9 - 12
Coal	2.0 - 3.0	*	12 - 16
Coke	2.5	*	9 - 12
Dyes	2.0	*	10
Fertilizer	2.0 - 3.5	1.8 - 2.0	8 - 10
Flint	2.5	*	*
Fly Ash	2.0	2.1 - 2.3	9 - 10
Foundry	*	*	8 - 12
Glass	2.5	*	*
Graphite	1.5 - 3.0	1.5 - 2.0	7 - 9
Gypsum	2.0 - 3.5	1.8 - 2.0	10 - 16
Iron Ore	2.0 - 3.5	*	11 - 12
Iron Oxide	2.0 - 3.0	1.5 - 2.0	8 - 16
Iron Sulfate	2.0 - 2.5	1.5 - 2.0	6 - 8
Lead Oxide	2.0 - 2.5	1.5 - 1.8	6 - 9
Leather	3.5 - 4.0	*	15 - 20
Lime	2.0 - 3.0	1.5 - 2.0	10 - 16
Limestone	2.0 - 3.3	*	8 - 12
Machining	3.0	*	16
Manganese	2.25	*	*
Metal Fumes	1.5	1.5 - 1.8	6 - 9
Metal Powders	2.0	*	9 - 10
Mica	2.25 - 3.3	1.8 - 2.0	9 - 11
Paint Pigments	2.0	*	*
Paper	3.5 - 4.0	*	10 - 12
Perchlorates	*	*	10
Plastics	2.0 - 3.0	*	7 - 10
Polyethylene	*	*	10
PVC	*	*	7
Resin	2.0	*	8 - 10
Silica	2.25 - 2.8	1.2 - 1.5	7 - 12
Silica Flour	2.0 - 2.5	*	*
Silicates	*	*	9 - 10
Silicon Carbide	*	*	10
Slate	2.5 - 4.0	*	12 - 14
Starch	2.25	*	*
Talc	2.25	*	*

* No information available.

4.9.4 Evaluation of Permit Application

Using Table 4-16, compare the results from this section and the data supplied by the permit applicant. The calculated values are based on the example case. As pointed out in the discussion on fabric filter design considerations, the basic design parameters are generally selected without the involved, analytical approach that characterizes many other control systems, such as an absorber system (Section 4.7). Therefore, in evaluating the reasonableness of any system specifications on a permit application, the reviewer's main task will be to examine each parameter in terms of its compatibility with the gas stream and particulate conditions and with the other selected parameters. The following questions should be asked:

1. Is the temperature of the emission stream entering the baghouse within 50° to 100°F above the stream dew point?
2. Is the selected fabric material compatible with the conditions of the emission stream; that is, temperature and composition (see Table 4-12)?

Table 4-15. Factors to Obtain Gross Cloth Area from Net Cloth Area (30)

Net Cloth Area, A_{nc} (ft^2)	Factor to Obtain Gross Cloth Area, A_{tc} (ft^2)
1 - 4,000	Multiply by 2
4,001 - 12,000	Multiply by 1.5
12,001 - 24,000	Multiply by 1.25
24,001 - 36,000	Multiply by 1.17
36,001 - 48,000	Multiply by 1.125
48,001 - 60,000	Multiply by 1.11
60,001 - 72,000	Multiply by 1.10
72,001 - 84,000	Multiply by 1.09
84,001 - 96,000	Multiply by 1.08
96,001 - 108,000	Multiply by 1.07
108,001 - 132,000	Multiply by 1.06
132,001 - 180,000	Multiply by 1.05
180,001 +	Multiply by 1.04

Example Case

Table 4-12 indicates that filter fabrics that can withstand the 400°F emission stream temperature are nylon aromatic (Nomex), fluorocarbon (Teflon), and fiberglass. Because there is a high potential for acid damage (i.e., a high SO_3 content), however, Nomex bags should not be considered. Because HAP's are present, only mechanical shaking or reverse air flow cleaning methods are advisable.

Using Table 4-14 for fly ash type dust, a low A/C ratio is expected for the two acceptable cleaning methods (2 to 2.3 ft/min). Because a fiberglass bag would provide the most protection during temperature surges, and because fiberglass bags may be less expensive, it may be the fabric of choice for an installation with these emissions characteristics. Fiberglass bags would require that reverse air cleaning be used. Teflon bags with mechanical shaking could also be a possibility. The documents that describe experience in certain industry applications support the choice of fiberglass bags with reverse air flow cleaning.(31,33)

Table 4-16. Comparison of Calculated Values and Values Supplied by the Permit Applicant for Fabric Filters

	Calculated Value (Example Case)[a]	Reported Value
Emission Stream Temp. Range[b]	365°-415°F	. . .
Selected Fabric Material	Fiberglass or Teflon	. . .
Baghouse Cleaning Method	Mechanical shaking or reverse air flow	. . .
A/C ratio $= \frac{Q_{e,a}}{A_{nc}}$	2-2.3 ft/min	. . .
Baghouse Configuration	Negative pressure	. . .

[a]Based on the municipal incinerator emission stream.
[b]See Section 3.3.1.

3. Is the baghouse cleaning method compatible with the selected fabric material and its construction; that is, material type and woven or felted construction (see Section 4.9.3.2 and Table 4-13)?
4. Will the selected cleaning mechanism provide the desired control?
5. Is the A/C ratio appropriate for the application; that is, type of dust and cleaning method used (see Table 4-14)?
6. Are the values provided for the gas flow rate, A/C ratio, and net cloth area consistent? The values can be checked with the following equation:

$$\text{A/C ratio} = \frac{Q_{e,a}}{A_{nc}} \quad (4.9\text{-}3)$$

where:
A/C ratio = air-to-cloth ratio, ft/min
$Q_{e,a}$ = emission stream flow rate at actual conditions, acfm
A_{nc} = net cloth area, ft^2

7. Is the baghouse configuration appropriate; that is, is it a negative-pressure baghouse?

A particular manufacturer/customer combination may employ somewhat different criteria in their selection of design parameters (such as lower annualized costs of operation at the expense of higher initial costs), and so a departure from the "rules-of-thumb" discussed here may still be compatible with achieving the needed high collection efficiencies. Further discussions with the permit applicant are recommended to evaluate the design assumptions and to reconcile any apparent discrepancies with usual practice.

4.9.5 Determination of Baghouse Operating Parameters

Many times, optimization of a fabric filter's collection efficiency occurs in the field after construction. The following discussion does not pertain to the preliminary design of a fabric filtration control system; however, the information presented should be helpful in achieving and maintaining the desired collection efficiency for the installed control system.

4.9.5.1 Collection Efficiency

A well designed fabric filter can achieve collection efficiencies in excess of 99 percent, although optimal performance of a fabric filter system may not occur for a number of cleaning cycles as the new filter material is "broken in." The fabric filter collection efficiency is related to the pressure drop across the system, component life, filter fabric, cleaning method and frequency, and A/C ratio. These factors should be reevaluated if fabric filter performance is less than permitted. Modifications to improve performance include changing the A/C ratio, using a different fabric, or replacing worn or leaking filter bags. Collection efficiency can be improved by decreasing the frequency of cleaning or allowing the system to operate over a greater pressure drop before cleaning is initiated.

4.9.5.2 System Pressure Drop

The pressure drop across the operating fabric filter system is a function of the difficulty with which the gas stream passes through the filter bags and accumulating dust cake, how heavy the dust deposit is prior to bag cleaning, how efficient cleaning is, and if the filter bags are plugged or blinded. Normally, the value of this parameter is set at about 3 to 4 inches of water, although pressure drops in excess of 10 inches have been used.(29) In actual operation, variations in pressure drop outside of the design range may be indicative of problems within the fabric filter system. Higher than expected pressure differentials may indicate: (1) an increase in gas stream volume; (2) blinding of the filter fabric; (3) hoppers full of dust, thus blocking the bags; and/or (4) inoperative cleaning mechanism. Lower than expected pressure differentials may indicate: (1) fan or motor problems, (2) broken or unclamped bags, (3) plugged inlet ducting or closed damper, and/or (4) leakage between sections of the baghouse.

As the dust cake builds up during filtration, both the collection efficiency and system pressure drop increase. As the pressure drop increases toward a maximum, the filter bags (or at least a group of the bags contained in one isolated compartment) must be cleaned to reduce the dust cake resistance. This cleaning must be timed and performed so as to accomplish the following: (1) to keep the pressure drop and thus operating costs, within reasonable limits; (2) to clean bags as gently and/or infrequently as possible to minimize bag wear and to maximize efficiency; and (3) to leave a sufficient dust layer on the bags to maintain filter efficiency and to keep the instantaneous A/C ratio immediately after cleaning from reaching excessive levels.

In practice, these various considerations are balanced using engineering judgment and field trial experience to optimize the total system operation. Changes in process or in fabric condition through fabric aging will cause a shift in the cleaning requirements of the system. This shift may require more frequent manual adjustments to the automatic control to achieve the minimum cleaning requirements.

4.10 Electrostatic Precipitators

Electrostatic precipitators (ESP's) use an electrostatic field to charge particulate matter contained in the gas stream. The charged particles then migrate to a grounded collecting surface. The collected particles are dislodged from the collector surface periodically by vibrating or rapping the collector surface, and subsequently collected in a hopper at the bottom of the ESP.

There are two basic types of ESP's: single stage and two stage.(32,37) In the single stage precipitator, which may be wet or dry, ionization and collection are combined, whereas in the two stage precipitator, ionization and collection are done in separate steps. Dry, single stage ESP's are the most common. Wet electrostatic precipitators, while not as common as dry ESP's, can be used to remove both solid and gaseous pollutants.

The most important variable considered in the design of an ESP is collection plate area; this assumes that the ESP is provided with an optimum level of secondary voltage. Collection plate area is a function of the desired collection efficiency, gas stream flow rate and particle drift velocity.(32,37,39,40) Other design details to be estimated by the vendor include (but are not limited to) expected secondary voltage and current, electrical sections alignment, and direction of gas flow. In this document, an approximate method to size an ESP is given.

Particle drift velocity is a complicated function of particle size, gas velocity, gas temperature, particle resistivity, particle agglomerization, and the physical and chemical properties of the particulate matter. The theoretical relationship of the drift velocity to the variables is discussed extensively in the literature.(32,37,39,40) Unfortunately, there are no empirical equations readily available to calculate drift velocity directly from these variables. Therefore, in determining drift velocity for a given emission stream, equipment vendors often rely upon historical data for similar streams and data established from pilot plant tests. Published information on drift velocity (based on design data for actual installations to represent typical gas characteristics) are available for several industrial emission streams.(37)

Appendix C.10 provides a worksheet to record the information obtained during the performance of the ESP design procedures.

4.10.1 Data Required

The data necessary to perform the design steps consist of the data characteristics previously compiled on the HAP Emission Stream Data Forms and the required HAP control as determined by the applicable regulations.

In the case of a permit review for an ESP, the following data should be supplied by the applicant: Reported collection efficiency, %

Reported drift velocity of particles, ft/sec
Reported collection plate area, ft^2

The design criteria and considerations discussed in this section will be used to evaluate the reasonableness of the applicant's proposed design.

> **Example Case**
> Electrostatic precipitation was one of the selected control techniques for the municipal incinerator stream. The pertinent data for these procedures are found on the HAP Emission Stream Data Form (see Figure 3-8).
>
> Flow rate, $Q_{e,a}$ = 110,000 acfm
> Emission stream temperature, T_e = 400°F
> Particulate content = 3.2 grains/scf - flyash
> Moisture content, M_e = 5% (vol)
> HAP content = 10% (mass) cadmium
> Drift velocity of particles, U_d = 0.3 ft/s
> Collection efficiency, CE = 99.9% mass

4.10.2 Pretreatment of the Emission Stream

As discussed in Section 3.3.1, the temperature of the emission streams should be within 50° to 100°F above the stream dew point. Procedures for determining the dew point of an emission stream are provided in Appendix B.1. If the emission stream temperature does not fall within the stated range, pretreatment (i.e., emission stream preheat or cooling) is necessary. (Methods of pretreatment are briefly discussed in Appendix B.11, reference 8.) The primary characteristics affecting ESP sizing are drift velocity of the particles and flow rate. Therefore, after selecting a temperature for the emission stream, the new stream flow rate must be calculated. The calculation method depends upon the type of pretreatment performed; use appropriate standard industrial equations. The use of pretreatment mechanical dust collectors may also be appropriate. If the emission stream contains an appreciable amount of large particles (20 to 30 μm), pretreatment with mechanical dust collectors is typically performed. (Appendix B.11, reference 8, further describes the use of mechanical dust collectors.)

4.10.3 ESP Design Variables

Estimating the collection plate area is the important aspect of sizing an ESP. A secondary consideration is the material of construction.

4.10.3.1 Collection Plate Area

Although precise specification of collection plate area is best left to the vendor, an approximate collection plate area can be calculated using the available drift velocity value for the gas stream.

As noted earlier, collection plate area is a function of the emission stream flow rate, the particulate drift velocity, and desired control efficiency. The Deutsch-Anderson equation relates these variables as follows:(32,37)

$$A_p = \frac{-Q_{e,a}}{60 \times u_d} \times \ln(1 - CE) \qquad (4.10\text{-}1)$$

where:

A_p = collection plate area, ft^2
$Q_{e,a}$ = emission stream flow rate at actual conditions as it enters the control device, acfm
U_d = drift velocity of particles, ft/s
CE = required collection efficiency, decimal fraction

Published data on drift velocities for a number of industrial applications are presented in Table 4-17. When unavailable, a drift velocity value for an industrial application can be obtained from an ESP vendor or from literature sources.(30) If no value for drift velocity is known, 0.30 ft/s for particles of "average" resistivity (approximately 10^7 to 2×10^{10} ohm-cm) and 0.10 ft/s for particles having a "high" resistivity (10^{11} to 10^{13} ohm-cm) can be used.(37)

Table 4-17. Typical Values for Drift Velocity for Various Particulate Matter Applications (37)

Application	Drift Velocity, ft/s
Pulverized Coal	0.33 to 0.44
Paper Mills	0.25
Open-hearth Furnace	0.19
Secondary Blast Furnace (80% foundry iron)	0.41
Gypsum	0.52 to 0.64
Hot Phosphorous	0.09
Acid Mist (H_2SO_4)	0.19 to 0.25
Acid Mist (TiO_2)	0.19 to 0.25
Flash Roaster	0.25
Multiple-hearth Roaster	0.26
Portland Cement (wet manufacturing)	0.33 to 0.37
Portland Cement (dry manufacturing)	0.19 to 0.23
Catalyst Dust	0.25
Gray-iron Cupola (iron-coke ratio = 10)	0.10 to 0.12

Particles with low resistivities impose special design considerations on an ESP. Such particles (resistivities from 10^4 to 10^7 ohm-cm) are difficult to collect in an ESP because the particles tend to lose their charge and drop off the collector plate and become reentrained in the gas stream. In such cases, specially designed collecting plates or coatings may be used to reduce reentrainment.(32,37) Particles with high resistivities also can cause ESP operating difficulties. High resistivity particles accumulate on the collection plates and insulate the collection plate, thus reducing the attraction between the particles and the collecting plate. In these cases, oversizing an ESP and more frequent cleaning or rapping of the collector plates are necessary. An alternative to a larger ESP is the use of conditioning agents to reduce the resistivity of the particles. Consult a vendor for advice concerning conditioning agents.

Example Case
Flow rate, $Q_{e,a}$ = 110,000 acfm
Drift velocity of particles, U_d = 0.30 ft/s
Collection efficiency, CE = 0.999
From the Deutsch-Anderson equation:
$A_p = \frac{-110{,}000 \text{ acfm } [\ln(1 - 0.999)]}{60 \times 0.30 \text{ ft/s}}$
A_p = 42,200 ft^2 of collection plate area

4.10.3.2 Materials of Construction

The most common material used in ESP construction is carbon steel. In cases where the gas stream contains high concentrations of SO_3 or where liquid-gas contact areas are involved, stainless steel may be required.(30,32,37,39,40) However, by keeping the emission stream temperature above the dew point and by insulating the ESP (the temperature drop across an insulated ESP should not exceed 20°F) the use of stainless steel should not be necessary.

4.10.4 Evaluation of Permit Application

Using Table 4-18, compare the results from this section and the data supplied by the permit applicant. The calculated values are based on the example. In evaluating the reasonableness of ESP design specifications in a permit application, the main task will be to examine each parameter in terms of its compatibility with the gas stream conditions.

If the applicant's collection plate area is less than the calculated area, the discrepancy will most likely be the selected drift velocity. Further discussions with the permit applicant are recommended to evaluate the design assumptions and to reconcile any apparent discrepancies.

Table 4-18. Comparison of Calculated Values and Values Supplied by the Permit Applicant for ESP's

	Calculated Value (Example Case)[a]	Reported Value
Drift velocity of particles, U_d	0.30 ft/s	. . .
Collection efficiency, CE	0.999	. . .
Collection plate area, A_p	42,200 ft^2	. . .

[a]Based on the municipal incinerator emission stream.

4.10.5 Determination of ESP Operating Parameters

Many times, optimization of an ESP's collection efficiency occurs in the field after construction. The following discussion does not pertain to the preliminary design of an ESP control system, however, the information presented should be helpful in achieving and maintaining the desired collection efficiency for the installed control system.

4.10.5.1 Electric Field Strength

Current in the form of ions from the charging electrodes actually charge the particles. Once the particles are charged, the electric field strength determines the amount of charge on the particles.

Field strength is based on voltage and distance between the collecting plates and electrodes.(32,37,39) ESP's are usually operated at the highest secondary voltage practicable with limited sparking to maximize collection efficiency. Sparking represents an instantaneous drop in voltage, collapse of the electrostatic field, and momentary cessation of particulate collection. Sparking varies with the density of the gas stream, material collected on the electrodes, and humidity and temperature of the gas stream. When automatic controls are used, ESP's usually operate with a small amount of sparking to ensure that the voltage is in the correct range and the field strength is maximized. Automatic voltage controls can control sparking to a specified sparking frequency (typically 50 to 150 sparks per minute per section of ESP).(37) As the spark rate increases, a greater percentage of the input power is wasted in the spark current. Consequently, less useful power is applied to the collecting electrode.

4.10.5.2 Cleaning Frequency and Intensity

Particles accumulating on the collecting plates must be removed periodically. In wet ESP's the liquid flowing down the collector surface removes the particles.(40) In dry ESP's, the particles are removed by vibrating or rapping the collector plates. For dry ESP's this is a critical step in the overall performance because improperly adjusted or operating rappers can cause reentrainment of collected particles or sparking due to excessive particulate buildup on the collection plates or discharge electrodes. In normal operation, dust buildup of 6 to 25 mm is allowed before rapping of a given intensity is initiated.(32) In this way, collected material falls off in large clumps that would not be reentrained. If rapping is initiated more frequently or if the intensity of rapping is lowered, the resulting smaller clumps of particulate matter are more likely to be reentrained, reducing the collection efficiency of the ESP. Optimal adjustment of the ESP can best be made by direct visual inspections through sight ports.

4.10.5.3 ESP Collection Efficiency

ESP collection efficiencies less than permitted can be the result of operational problems, mechanical troubles, or improper design. Typical operational problems include improper electrical settings, badly adjusted rappers, full or nearly full dust hoppers, and process upsets. Mechanical difficulties typically are the result of electrode misalignment or excessive dust buildup on the electrodes. Basic design problems include undersized equipment, reentrainment, or high resistivity particles. The permit applicant should carefully examine each of these items if the ESP is emitting particulate emissions from his facility that are in excess of permitted levels.

4.11 Venturi Scrubbers

Venturi scrubbers are designed to serve as a control device for applications requiring very high collection efficiencies of particles generally between 0.5 to 5.0 μm in diameter. They employ gradually converging and then diverging sections to clean an incoming gaseous stream. The section connecting the converging and diverging sections of the scrubber is called the throat. In general, the longer the throat, the higher the collection efficiency at a given pressure drop, provided the throat is not so long that frictional losses become significant.(42) Typically, a liquid (usually water) is introduced upstream of the throat and flows down the converging sides into the throat where it is atomized by the gaseous stream; this method is called the "wetted approach." Alternatively, the liquid can be injected into the throat itself by use of nozzles directed at the throat; this approach is called the "nonwetted approach."(42) The nonwetted approach works well when a gas is already close to saturation; however, this method requires that the liquid be free of particles that could clog the nozzles. Where inlet gases are hot and a significant amount of liquid needs to be evaporated, the wetted approach is preferred.

Once the liquid is atomized, it begins to collect particles from the gas impacting into the liquid as a result of the difference in velocities of the gas stream and the atomized droplets. As the mixture decelerates in the expanding section, further impaction occurs causing the droplets to agglomerate. Once the particles have been trapped by the liquid, a separator (e.g. cyclone, demisters, swirl vanes) can readily remove the scrubbing liquid from the cleaned gas stream.

Appendix C.11 provides a worksheet to record the information obtained during the performance of the venturi scrubber design procedures.

4.11.1 Data Required

The data necessary to perform the design steps consist of the HAP emission stream characteristics previously compiled on the HAP Emission Stream Data Forms, and the required HAP control as determined by the applicable regulations.

In the case of a permit review for a venturi scrubber, the following data should be supplied by the applicant:

Reported pressure drop across venturi, in H_2O
Performance curve applicable to the venturi scrubber
Reported collection efficiency, %

Example Case

A venturi scrubber was one of the selected control techniques for the municipal incinerator emission stream. The pertinent data for these procedures are found on the HAP Emission Stream Data Form.

Flow rate $Q_{e,a}$ = 110,000 acfm
Temperature, T_e = 400°F
Moisture content, M_e = 5% vol
Required collection efficiency, CE = 99.9%
Particle mean diameter, D_p = 1.0 μm
Particulate content = 3.2 grams/scf flyash
HAP content = 10% (mass) cadmium

4.11.2 Pretreatment of the Emission Stream

As discussed in Section 3.3.1, the temperature of the emission stream should be within 50° to 100°F above the stream dew point. Procedures for determining the dew point of an emission stream are provided in Appendix B.1. If the emission stream temperature does not fall within the stated range, pretreatment (i.e., emission stream preheating or cooling) is necessary. (Methods of pretreatment are briefly discussed in reference 32 and Appendix B.11, reference 8.) If pretreatment is performed, the emission stream characteristics will be altered. The primary characteristic affecting venturi scrubber design is the saturated gas flow rate ($Q_{e,s}$), a function of the emission stream temperature (T_e) and flow rate at actual conditions ($Q_{e,a}$).(42) Thus, if the temperature of the emission stream changes, thus changing the actual flow rate, the saturated gas flow rate must be based on the new actual flow rate. The calculation method depends upon the type of pretreatment performed; use appropriate standard industrial equations. The use of pretreatment mechanical dust collectors may also be appropriate, particularly if a "nonwetted" venturi scrubber is used.

4.11.3 Venturi Scrubber Design Variables

To design a venturi scrubber, any one of three paths may be chosen: (1) rely on previous experience with an analogous application, which is best for plants lacking effluent data; (2) test a scrubber on the source itself; or (3) collect sufficient data about source stream characteristics, such as particle size distribution, flow rate, and temperature, to utilize existing "performance curves" for a given venturi scrubber. This section is concerned with the third path. Thus, the most important consideration becomes the pressure drop across the venturi. A secondary consideration is materials of construction.

4.11.3.1 Venturi Scrubber Pressure Drop

Performance curves are typically logarithmic plots relating venturi collection efficiency, pressure drop, and particle size.(23,32,43,44) Collection (control) efficiency is usually plotted versus pressure drop across the venturi (ΔPv) for a particle mean diameter (D_p). Figure 4-20 is a plot of venturi scrubber pressure drops for a given collection efficiency and particle mean diameter for venturi scrubbers manufactured by a specific vendor. Thus, if the particle mean diameter for an emission stream and required collection efficiency is known, the pressure drop across the venturi can be estimated. Figure 4-20 is representative of plots likely to be used by vendors, and does not necessarily represent characteristics for all venturi scrubbers.

Estimating the pressure drop gives an indication of whether a venturi scrubber is a feasible control device for a given stream. Venturi scrubbers are used in applications where pressure drops of between 10 and 80 inches water gauge occur across the venturi. Venturi scrubbers can operate at pressure drops higher than 80 inches; however, in general, a pressure drop exceeding 80 inches H_2O indicates that a venturi scrubber will have difficulty collecting the particles.(42) Therefore, if the pressure drop indicated on the performance curve is greater than 80 inches H_2O, assume that the venturi scrubber cannot accomplish the desired control efficiency.

Table 4-19 lists typical pressure drops for venturi scrubbers for a variety of applications. The pressure drops are listed to provide general guidance for typical values that occur in industry. The values are not meant to supersede any specific information known, and given application may have a pressure drop outside those listed in Table 4-19.

Figure 4-20. Venturi scrubber collection efficiencies.

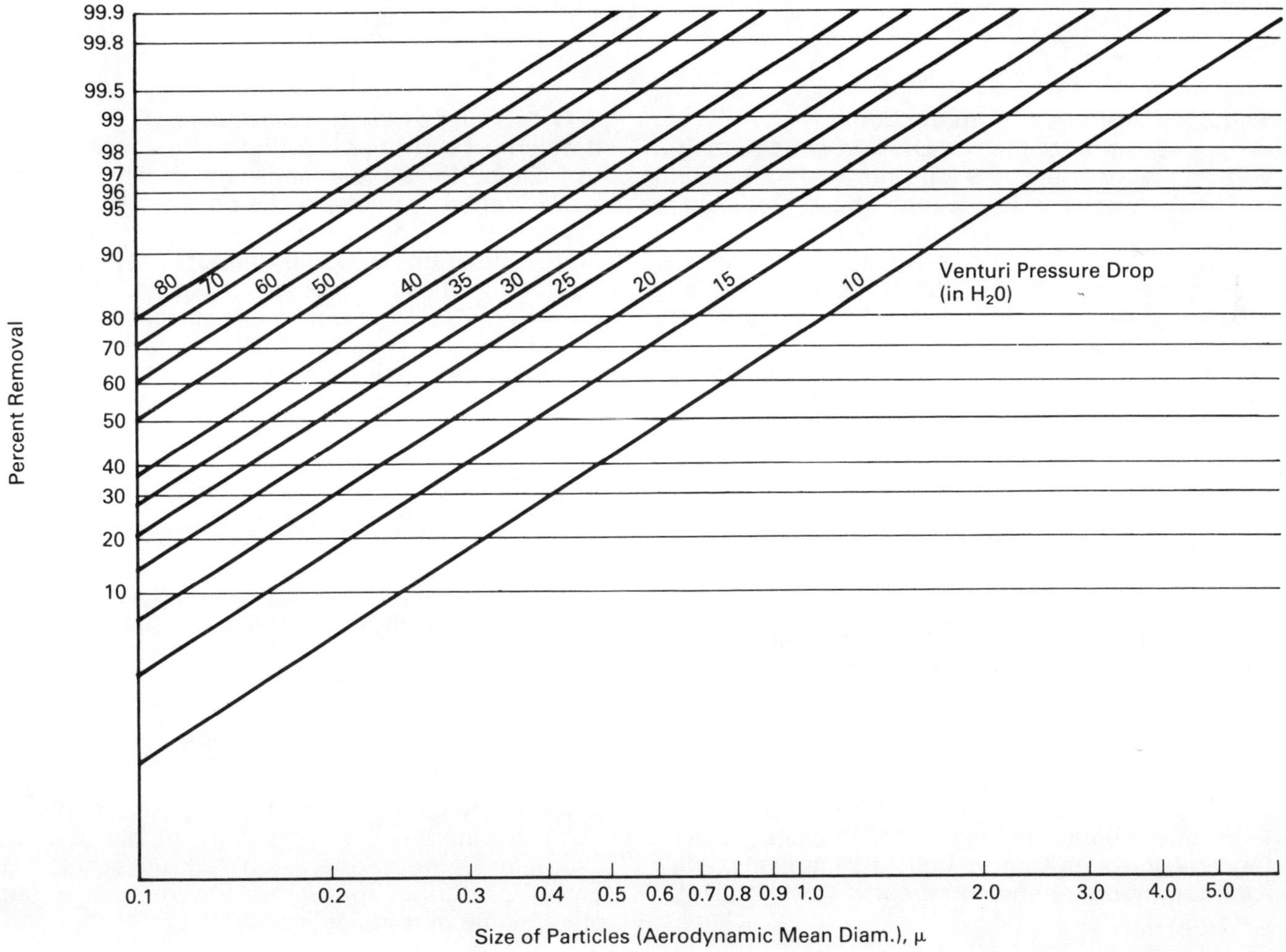

Table 4-19. Pressure Drops for Typical Venturi Scrubber Applications (42)

Application	Pressure drop (in H_2O)	Application	Pressure drop (in H_2O)
Boilers		Iron and Steel	
Pulverized coal	15-40	Cupolas	30-50
Stoker coal	10-12	Arc furnaces	30-50
Bark	6-10	BOF's	40-60
Combination	10-15	Sand systems	10
Recovery	30-40	Coke ovens	10
Incinerators		Blast furnaces	20-30
Sewage sludge	18-20	Open hearths	20-30
Liquid waste	50-55	Nonferrous metals	
Solid waste		Zinc smelters	20-50
Municipal	10-20	Copper and brass	
Pathological	10-20	smelters	20-50
Hospital	10-20	Sinter operations	20
Kilns		Aluminum reduction	50
Lime	15-25	Phosphorus	
Soda Ash	20-40	Phosphoric acid	
Potassium chloride	30	Wet process	10-30
Coal Processing		Furnace grade	40-80
Dryers	25	Asphalt	
Crushers	6-20	Batch plants—dryer	10-15
Dryers		Transfer points	6-10
General spray	20-60	Glass	
Food spray	20-30	Container	25-60
Fluid bed	20-30	Plate	25-60
Mining		Borosilicate	30-60
Crushers	6-20	Cement	
Screens	6-20	Wet process kiln	10-15
Transfer points	6-20	Transfer points	6-12

[a]Source: Reference 1.

4.11.3.2 Materials of Construction

Proper selection of the materials in constructing a venturi scrubber ensures long-term operation with minimal downtime for repair. The materials are generally chosen based on the corrosive or erosive nature of the emission stream, and to a lesser degree, the temperature of the gas stream. For any given application, a vendor should be contacted to ensure correct selection of materials. A venturi scrubber will generally be constructed of either carbon or stainless steel or a nickel alloy; it may also be lined with another material (e.g., ceramics). Table 4-20 lists materials of construction for various industries and is intended to serve as a general guide rather than a definitive statement on the types of materials used in industry.

> **Example Case**
>
> The required collection efficiency is 99.9% and the particle mean diameter in the municipal waste incinerator emission stream is estimated to be 1.0 μm; therefore:
>
> $\Delta P_v = 47$ in H_2O (Figure 4-20)
>
> Since the estimated venturi pressure drop value of ΔP_v is not greater than 80 in H_2O, this venturi scrubber should be able to accomplish the desired control efficiency. Table 4-20 indicates the venturi scrubber should be constructed of 316L stainless steel.

4.11.4 Sizing of Venturi Scrubbers

If a venturi scrubber is found to be a feasible control choice for a given emission stream, it is then sized. Venturi scrubbers can be sized using either the flowrate at inlet conditions ($Q_{e,a}$) or the saturated gas flowrate ($Q_{e,s}$).(45) Vendors may use either parameter; the cost data presented in Chapter 5 are based on $Q_{e,a}$. However, more current cost curves based on $Q_{e,s}$ may be available; therefore, $Q_{e,s}$ should be calculated. A psychrometric chart (Figure 4-21) can be used to determine the saturated gas temperature ($T_{e,s}$), and $Q_{e,s}$ can then be calculated using the following formula:

$$Q_{e,s} = Q_{e,a} \times (T_{e,s} + 460)/(T_e + 460) \qquad (4.11\text{-}1)$$

where:

$Q_{e,s}$ = saturated emission stream flow rate, acfm

$T_{e,s}$ = temperature of the saturated emission stream, °F

4.11.5 Evaluation of Permit Application

Using Table 4-21, compare the results of this section and the data supplied by the permit applicant. The calculated values in the table are based on the example. Compare the estimated ΔP_v and the reported pressure drop across the venturi, as supplied by the permit applicant.

Table 4-20. Construction Materials for Typical Venturi Scrubber Applications[a]

Application	Construction Material	Application	Construction Material
Boilers		Iron and Steel	
Pulverized coal	316L stainless steel	Cupolas	304-316L stainless steel
Stoker coal	316L stainless steel	Arc furnaces	316L stainless steel
Bark	Carbon steel	BOF's	Carbon steel (ceramic lined)
Combination	316L stainless steel	Sand systems	Carbon steel
Recovery	Carbon steel or 316L stainless steel	Coke ovens	Carbon steel
Incinerators		Blast furnaces	Carbon steel (ceramic lined)
Sewage sludge	316L stainless steel	Open hearths	Carbon steel (ceramic lined)
Liquid waste	High nickel alloy	Nonferrous Metals	
Solid waste		Zinc smelters	Stainless steel or high nickel
Municipal	316L stainless steel	Copper and brass smelters	Stainless steel or high nickel
Pathological	316L stainless steel	Sinter operations	Stainless steel or high nickel
Hospital	High nickel alloy	Aluminum reduction	High nickel
Kilns		Phosphorus	
Lime	Carbon steel or stainless steel	Phosphoric acid	
Soda Ash	Carbon steel or stainless steel	Wet process	316L stainless steel
Potassium chloride	Carbon steel or stainless steel	Furnace grade	316L stainless steel
Coal Processing		Asphalt	
Dryers	304 stainless steel or 316L stainless steel	Batch plants—dryer	Stainless steel
Crushers	Carbon steel	Transfer points	Carbon steel
Dryers		Glass	
General spray dryer	Carbon steel or stainless steel	Container	Stainless steel
Food spray dryer	Food-grade stainless steel	Plate	Stainless steel
Fluid bed dryer	Carbon steel or stainless steel	Borosilicate	Stainless steel
Mining		Cement	
Crushers	Carbon steel	Wet process kiln	Carbon steel or stainless steel
Screens	Carbon steel	Transfer points	Carbon steel
Transfer points	Carbon steel		

Example Case

Determine saturated gas flow rate:
Emission stream flow rate, $Q_{e,a}$ = 110,000 acfm
Moisture content, M_e = 5% vol
Emission stream temperature, T_e = 400°F

Convert M_e to units of lb H_2O/lb dry air, decimal fraction:
$(M_e/100)\ (18/29) = (5/100)\ (18/29) =$
0.031 lb H_2O/lb dry air
$T_{e,s}$ = 127°F (Figure 4-21)
$Q_{e,s} = (110{,}000) \times (127 + 460)/(400 + 460) =$
75,000 acfm

If the estimated and reported values differ, the differences may be due to the applicant's use of another performance chart, or a discrepancy between the required and reported collection efficiencies. Discuss the details of the design and operation of the system with the applicant. If there are no differences between the estimated and reported values for ΔP_v, the design and operation of the system can be considered appropriate based on the assumptions employed in this handbook.

Table 4-21. Comparison of Calculated Values and Values Supplied by the Permit Applicant for Venturi Scrubbers

	Calculated Value (Example Case)[a]	Reported Value
Particle mean diameter, D_p	1.0 μm	. . .
Collection efficiency, CE	0.999	. . .
Pressure drop across venturi, ΔP_v	47 in H_2O	. . .

[a]Based on the municipal incinerator emission stream.

Figure 4-21. Psychrometric chart, temperature range 0° - 500°F, 29.92 in Hg pressure.

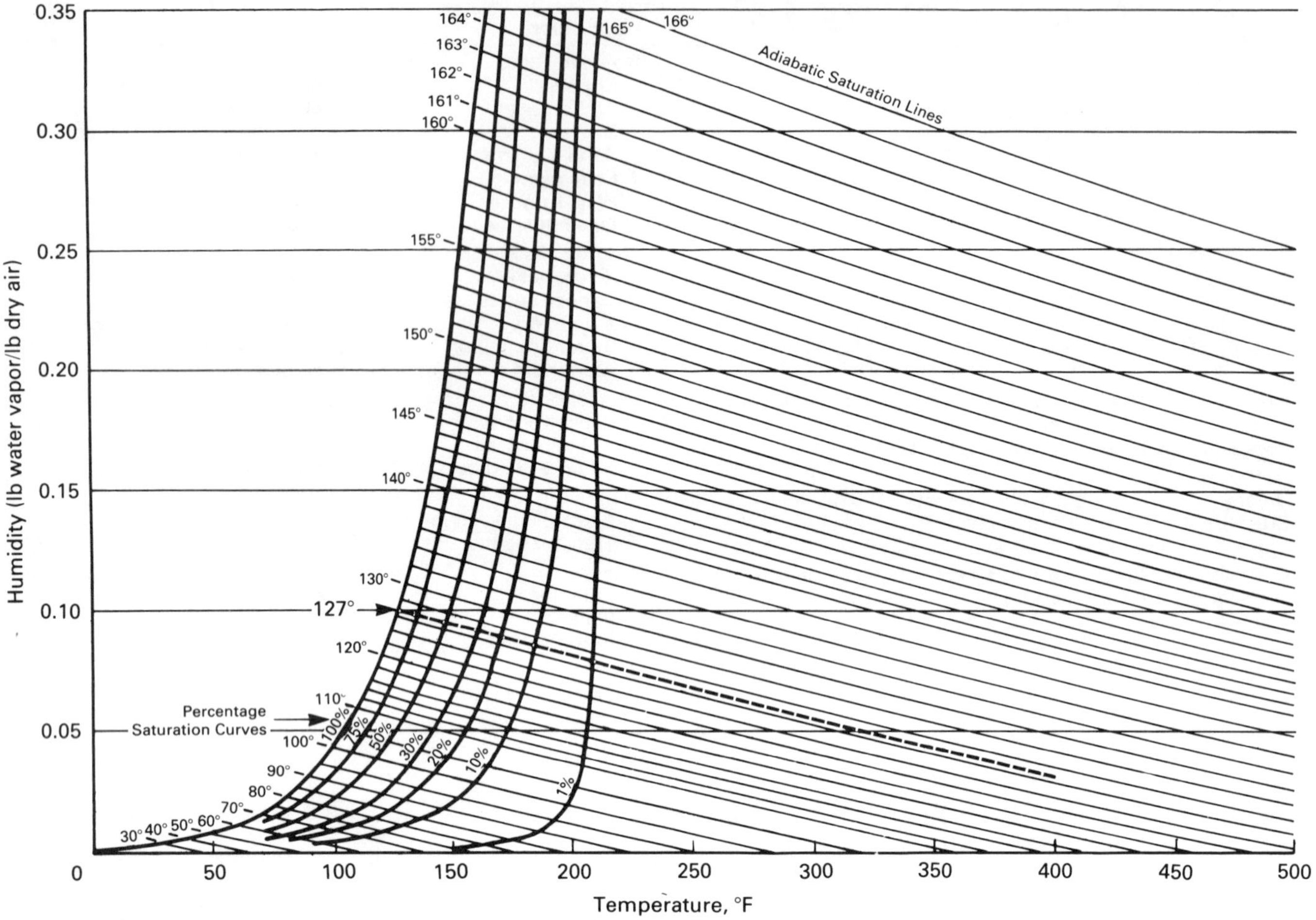

4.12 References

1. U.S. EPA. *Organic Chemical Manufacturing. Volume 4: Combustion Control Devices*. EPA-450/3-80-026. December 1980.

2. U.S. EPA. *Reactor Processes in Synthetic Organic Chemical Manufacturing Industry — Background Information for Proposed Standards*. Draft EIS. October 1984.

3. Hougen, O.A., K. M. Watson, and R. A. Ragatz. *Chemical Process Principles Part I: Material and Energy Balances*. Asia Publishing House. Bombay. 1962.

4. U.S. EPA. *Parametric Evaluation of VOC/HAP Destruction Via Catalytic Incineration*. EPA-600/2-85-041. April 1985.

5. U.S. EPA. *Control of Volatile Organic Emissions from Existing Stationary Sources — Volume I: Control Methods for Surface Coating Operations*. EPA-450/2-76-028. November 1976.

6. U.S. EPA. *Guideline Series. Control of Volatile Organic Compound Emissions From Manufacture of High-Density Polyethylene, Polypropylene, and Polystyrene Resins*. EPA-450/3-83-008. November 1983.

7. U.S. EPA. *Afterburner Systems Study*. EPA-R2-72-062. August 1972.

8. U.S. EPA. *Evaluation of Control Technologies for Hazardous Air Pollutants — Appendices*. EPA-600/7-86-009b (NTIS PB 86-167/038/AS). October 1985.

9. *Federal Register*. Volume 50. April 16, 1985. pp. 14941-14945.

10. U.S. EPA. *Evaluation of the Efficiency of Industrial Flares: Test Results*. EPA-600/2-84-095. May 1984.

11. Chandrasekhar, R., and E. Poulin. *Control of Hydrocarbon Emissions From Cotton and Synthetic Textile Finishing Plants*. EPA Contract No. 68-02-3134. May 1983.

12. U.S. EPA. *Organic Chemical Manufacturing Volume 5: Adsorption, Condensation, and Absorption Devices*. EPA-450/3-80-027. December 1980.

13. Parmele, C. S., W. L. O'Connell, and H. S. Basdekis. Vapor-phase Adsorption Cuts Pollution, Recovers Solvents. *Chemical Engineering*. December 31, 1979. pp. 58-70.

14. Calgon Corporation, Pittsburgh, Pennsylvania. In-house data.

15. Tomany, J. P. *Air Pollution: The Emissions, the Regulations, and the Controls*. American Elsevier Publishing Company, Inc. New York.

16. Manzone, R. R., and D. W. Oakes. Profitably Recycling Solvents From Process Systems. *Pollution Engineering*. October 1973. pp. 23-24.

17. Vatavuk, M. W., and R. B. Neveril. Part XIV: Costs of Carbon Adsorbers. *Chemical Engineering*. January 24, 1983. pp. 131-132.

18. *Chemical Engineer's Handbook*. Perry, R. H., and C. H. Chilton, eds. Fifth Edition. McGraw-Hill Book Company. New York. 1973.

19. Smith, J. M., and H. C. Van Ness. *Introduction to Chemical Engineering Thermodynamics*. Second Edition. McGraw-Hill Book Company, Inc. and Kogakusha Company, Ltd. Tokyo. 1959.

20. Treybal, R. E. *Mass-Transfer Operations*. Third Edition. McGraw-Hill Book Company. New York. 1980.

21. Buonicore, A. J., and L. Theodore. *Industrial Control Equipment for Gaseous Pollutants. Volume I*. CRC Press, Inc. Cleveland, Ohio. 1975.

22. Kohl, A., and F. Riesenfeld. *Gas Purification*. Second Edition. Gulf Publishing Company. Houston, Texas. 1974.

23. U.S. EPA. *Wet Scrubber System Study, Volume I: Scrubber Handbook*. EPA-R2-72-118a. August 1972.

24. Vatavuk, W. M., and R. B. Neveril. Part XIII. Costs of Gas Absorbers. *Chemical Engineering*. October 4, 1982. pp. 135-136.

25. Kern, D. Q. *Process Heat Transfer*. McGraw-Hill Book Company, Inc. and Kogakusha Company, Ltd. Tokyo. 1950.

26. Ludwig, E. E. *Volume III. Applied Process Design for Chemical and Petrochemical Plants*. Gulf Publishing Company. Houston, Texas. 1965.

27. *Lange's Handbook of Chemistry*. Dean, J. A., ed. Twelfth Edition. McGraw-Hill Book Company. New York. 1979.

28. Siebert, P. C. *Handbook on Fabric Filtration*. IIT Research Institute. Chicago. April 1977.

29. U.S. EPA. *Handbook of Fabric Filter Technology, Volume I: Fabric Filter Systems Study.* APTD 0690. December 1970.

30. U.S. EPA. *Capital and Operating Costs of Selected Air Polution Control Systems.* EPA-450/5-80-002. December 1978.

31. *The Fabric Filter Manual.* The McIlvaine Company. Northbrook, Illinois. 1975.

32. *Environmental Engineers' Handbook, Volume II: Air Pollution*. Liptak, B. G., ed. Chilton Book Company. Radnor, Pennsylvania. 1974.

33. U.S. EPA. *Control Techniques for Particulate Emissions from Stationary Sources — Volume 2*. EPA-450/3-81-005b. September 1982.

34. U.S. EPA. *Control Techniques for Particulate Emissions From Stationary Sources — Volume I*. EPA-450/3-81-005a. September 1982.

35. Strauss, W. *Industrial Gas Cleaning, 2nd Edition*. Pergamon Press, Oxford, England. 1975.

36. U.S. EPA. *Procedures Manual for Fabric Filter Evaluation*. EPA-600/7-78-113. June 1978.

37. U.S. EPA. *Air Pollution Engineering Manual*. AP-40. May 1973.

38. U.S. EPA. *Particulate Control Highlights: Research on Fabric Filtration Technology*. EPA-600/8-78-005d. June 1978.

39. U.S. EPA. *A Manual of Electrostatic Precipitator Technology, Part 1 — Fundamentals*. APTD 0610. 1970.

40. *Perry's Chemical Engineers' Handbook*. Perry, R. H., and D. Green, eds. Sixth Edition. McGraw-Hill Book Company. New York. 1984.

41. *The Electrostatic Precipitator Manual*. The McIlvaine Company. Northbrook, Illinois. 1975.

42. *Air Pollution Control and Design Handbook: Part 2*. Cheremisinoff, P. N., and R. A. Young, eds. Marcel Dekker, Inc. New York. 1977.

43. U.S. EPA. *Wet Scrubber Performance Model*. EPA 600/2-77-127. August 1977.

44. U.S. EPA. *TI-59 Programmable Calculator Programs for Opacity, Venturi Scrubbers and Electrostatic Precipitators*. EPA-600/8-80-024. May 1980.

45. U.S. EPA. *The Cost Digest. Cost Summaries of Selected Environmental Control Technologies*. EPA-600/8-84-010. October 1984.

Chapter 5
Cost Estimation Procedure

5.1 Objective

This chapter provides generalized procedures for estimating capital and annualized costs (June 1985 dollars) for a given add-on HAP control system. (Note: Calculation of the cost of HAP waste disposal is outside the scope of this handbook; however, this cost must be included in any rigorous control cost estimation.) The procedures are presented in a step-by-step format and illustrated at each step with cost calculations pertaining to the thermal incinerator system example discussed in Section 4.2. Blank standard cost calculation worksheets are provided in Appendix C.12.

Only the process HAP control systems presented in Chapter 4 are discussed in this chapter. The cost of fugitive emission controls are outside the scope of this handbook; however, an EPA report entitled *Identification, Assessment, and Control of Fugitive Particulate Emissions* (1) can be used for estimating costs of fugitive emission controls.

5.2 Total Capital Cost

In this handbook the total capital cost includes only manufacturing area costs; therefore, it excludes offsite costs. The total capital cost of a control system is the sum of direct costs, indirect costs, and contingency costs. Direct costs include the total purchased equipment cost (i.e., the major equipment purchased cost plus the auxiliary equipment purchased cost), instrumentation and controls, freight and taxes, and installation costs (i.e., foundation and supports, erection and handling, electrical, piping, insulation, and painting). (Note: The summation of the total purchased equipment cost, the cost of instrumentation and controls, and freight and taxes is defined as the total purchased cost.) Indirect costs consist of inhouse engineering design and supervision costs, architect and engineering contractor expenses, contractor fees, construction expenses, and preliminary testing costs. An example of contingency costs are penalties incurred for failure to meet completion dates or performance specifications.

The capital cost estimation procedure presented in this handbook is for a "factored" or "study" estimate. Usual reliability for a study type estimate is ± 30 percent. To determine the total capital cost by a factored cost estimate, a reliable estimate of the total purchased cost is calculated and predetermined factors are applied to determine all other capital cost elements. Therefore, the procedure to estimate the total capital cost is as follows: (1) obtain the total purchased equipment cost by estimating the purchased cost of major and auxiliary equipment; (2) estimate the cost of instrumentation and controls plus freight and taxes as a percentage of the total purchased equipment cost; (3) estimate the total purchased cost by adding (1) and (2) above; and (4) estimate total capital cost by applying a predetermined cost factor to the total purchased cost.

5.2.1 Estimation of Major Equipment Purchased Cost

The major equipment purchased cost (i.e., the cost of the major components that comprise the control system) is related to a specific equipment design parameter and can be expressed either analytically or graphically. Table 5-1 presents a list of the design parameters needed for costing the HAP control equipment, and it identifies the figure that presents the applicable purchased cost curve. Gathering current costs from vendors was beyond the scope of this project and thus, necessitated use of dated cost data compiled by others. In general,

Table 5-1. Identification of Design Parameters and Cost Curves for Major Equipment

Control Equipment	Design Parameter[a]	Cost Curve Figure No.
Thermal Incinerator	V_c	5-1
Heat Exchanger	A	5-2
Catalytic Incinerator	Q_{fg}	5-3
Carbon Adsorber	C_{req}	5-4[b] 5-5[c]
Absorber	Wt_{col} D_{col}	5-6 5-7
Condenser	A_{con} Ref	5-8 5-9
Fabric Filter	A_{nc}	5-10
Electrostatic Precipitator	A_p	5-11
Venturi Scrubber	$Q_{e,a}$	5-12 5-13 5-14

[a] See Nomenclature for definitions of variables.
[b] Packaged carbon adsorbers.
[c] Custom carbon adsorbers.

cost estimates should not be escalated beyond 5 years. If more recent cost data are available, they should be substituted for the cost curve data presented. These cost curves should not be extrapolated beyond their range. The cost data presented in these figures were obtained from cost information published in EPA reports.(2,3) To escalate the cost data to June 1985 dollars, multiply the cost estimate by the ratio of the Chemical Engineering Fabricated Equipment (FE) cost indices for June 1985 and the date of the cost data. For example, if a cost is given in December 1977 dollars, it is converted to June 1985 dollars using a factor of 1.49 (336.0 [June 1985]/226.2 [Dec. 1977]). Table 5-2 presents the monthly FE cost indices from December 1977 through June 1986.

Using the specific value for the design variable, obtain purchased costs from the specific cost curve for each major control system component. Presented below are brief descriptions of the equipment costs included in each HAP control cost curve.

The cost curve for *thermal incinerators* (Figure 5-1) includes the fan plus instrumentation and control costs, in addition to the major equipment purchased cost. If the HAP control system includes a heat exchanger, its cost (Figure 5-2) is part of the major equipment purchased cost and thus, must be added. The remaining auxiliary equipment (ductwork and stack) purchased costs and costs of freight and taxes must be added to obtain the total purchased cost.

The cost curve for *catalytic incinerators* (Figure 5-3) provides the cost of an incinerator less catalyst. Catalyst costs (Table 5-3) and the cost of a heat exchanger, if applicable, (Fig. 5-2) must be added to obtain the major equipment purchased cost. All auxiliary equipment (ductwork, fan, and stack) purchased costs, the cost of instrumentation and controls, and freight and taxes must be added to obtain the total purchased cost.

Figure 5-1. Prices for thermal incinerators, including fan and motor, and instrumentation and controls costs.(2)

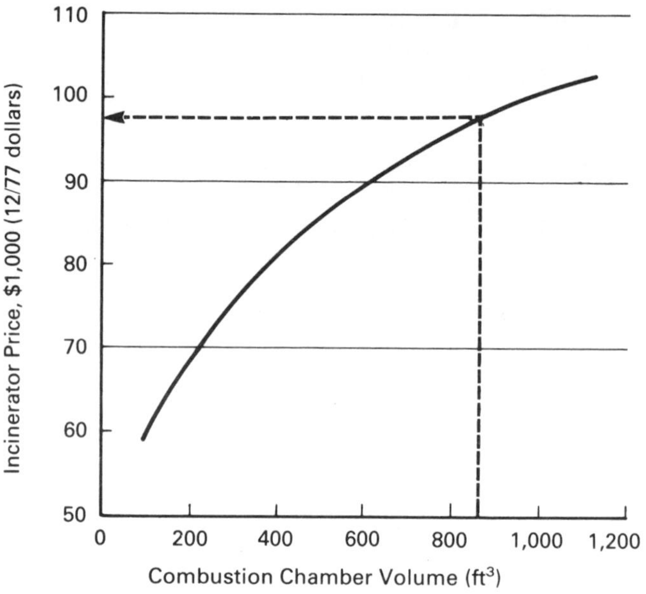

Table 5-2. Chemical Engineering Fabricated Equipment (FE) Cost Indices[a]

Date	FE	Date	FE	Date	FE	Date	FE
Dec. 1977	226.2	Jan. 1980	273.8	Mar. 1982	324.1	May 1984	334.6
		Feb. 1980	276.9	Apr. 1982	327.8	June 1984	333.8
Jan. 1978	226.6	Mar. 1980	277.7	May 1982	329.1	July 1984	335.4
Feb. 1978	233.0	Apr. 1980	289.3	June 1982	327.5	Aug. 1984	335.1
Mar. 1978	233.6	May 1980	290.9	July 1982	327.1	Sept. 1984	335.9
Apr. 1978	237.1	June 1980	291.3	Aug. 1982	326.2	Oct. 1984	335.0
May 1978	237.3	July 1980	296.7	Sept. 1982	326.7	Nov. 1984	335.4
June 1978	237.4	Aug. 1980	297.3	Oct. 1982	325.8	Dec. 1984	336.5
July 1978	238.6	Sept. 1980	298.1	Nov. 1982	324.8		
Aug. 1978	243.3	Oct. 1980	301.2	Dec. 1982	325.1	Jan. 1985	336.9
Sept. 1978	243.2	Nov. 1980	302.5			Feb. 1985	336.5
Oct. 1978	243.8	Dec. 1980	304.0	Jan. 1983	324.4	Mar. 1985	336.6
Nov. 1978	244.1			Feb. 1983	327.6	Apr. 1985	338.0
Dec. 1978	245.2	Jan. 1981	305.9	Mar. 1983	326.8	May 1985	336.0
		Feb. 1981	307.1	Apr. 1983	326.6	June 1985	336.2
Jan. 1979	245.2	Mar. 1981	314.7	May 1983	327.1	July 1985	336.4
Feb. 1979	252.5	Apr. 1981	321.9	June 1983	327.3	Aug. 1985	335.3
Mar. 1979	253.1	May 1981	321.6	July 1983	327.0	Sept. 1985	335.9
Apr. 1979	253.7	June 1981	322.9	Aug. 1983	327.1	Oct. 1985	335.4
May 1979	258.3	July 1981	325.6	Sept. 1983	328.0	Nov. 1985	335.7
June 1979	259.9	Aug. 1981	325.7	Oct. 1983	327.8	Dec. 1985	336.8
July 1979	262.6	Sept. 1981	326.7	Nov. 1983	328.9		
Aug. 1979	264.2	Oct. 1981	330.8	Dec. 1983	330.1	Jan. 1986	332.5
Sept. 1979	266.6	Nov. 1981	329.4			Feb. 1986	319.2
Oct. 1979	271.6	Dec. 1981	328.9	Jan. 1984	331.5	Mar. 1986	317.0
Nov. 1979	272.6			Feb. 1984	333.0	Apr. 1986	310.6
Dec. 1979	273.7	Jan. 1982	324.5	Mar. 1984	332.9	May. 1986	310.9
		Feb. 1982	323.4	Apr. 1984	333.8	June 1986[b]	310.8

[a] Source: *Chemical Engineering*, McGraw-Hill Publications.
[b] Preliminary.

Figure 5-2. Prices for thermal oxidation recuperative heat exchangers.(2)

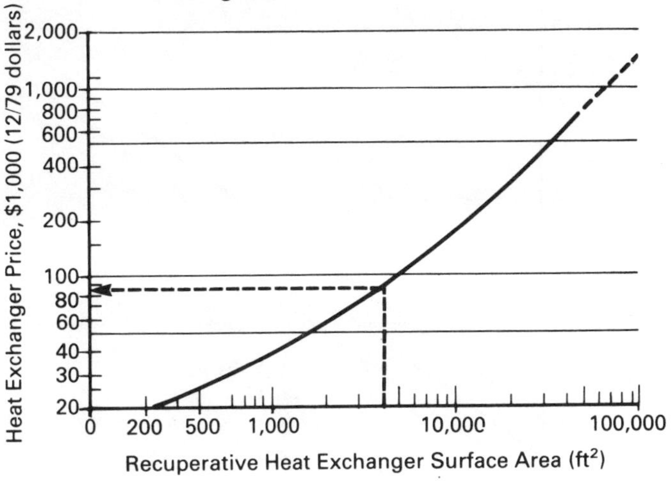

Figure 5-3. Prices for catalytic incinerators, less catalyst.(4)

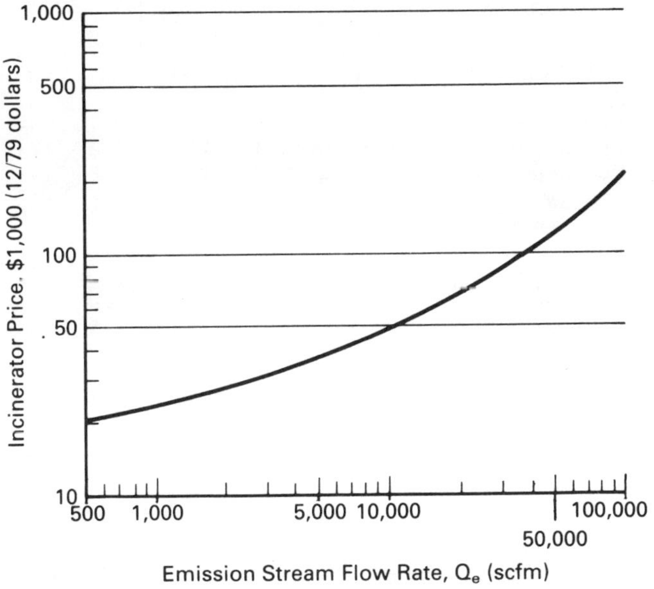

Table 5-3. Unit Costs for Various Materials (6/85 dollars)

Chemical	Cost
Refrigerant (ethylene glycol)	$0.31/lb (8)
Activated Carbon	$1.92/lb (9)
Catalyst (platinum-based)	$2,750/$ft^3$ (10)

Example Case

The example thermal incinerator system case (see Section 4.2) consists of an incinerator with a combustion chamber volume (V_c) of approximately 860 ft^3 and a primary heat exchanger with a surface area (A) of approximately 4,200 ft^2. From the cost data presented in Figures 5-1 and 5-2, June 1985 cost estimates are obtained as follows:

(a) Incinerator plus instrumentation and control costs—
$98,000 x (336.2/226.2) = $145,700
(Note: 12/77 dollars escalated to reflect 6/85 dollars.)

(b) Heat exchanger cost—
$85,000 x (336.2/273.7) = $104,400
(Note: 12/79 dollars escalated to reflect 6/85 dollars.)

Two cost curves are presented for *carbon adsorbers*: Figure 5-4 for packaged carbon adsorbers and Figure 5-5 for custom carbon adsorbers. The cost curve for packaged carbon adsorbers includes the fan plus instrumentation and control costs, in addition to the major equipment purchased cost. The cost of the remaining auxiliary equipment (ductwork and stack), as well as, costs of freight and taxes must be added to obtain the total purchased cost. The cost curve for custom carbon adsorbers does not include the cost of carbon (part of the major equipment purchased cost), however, it does include the cost of instrumentation and controls; the cost of carbon is obtained from Table 5-3.

Figure 5-4. Prices for carbon adsorber packages.(5)

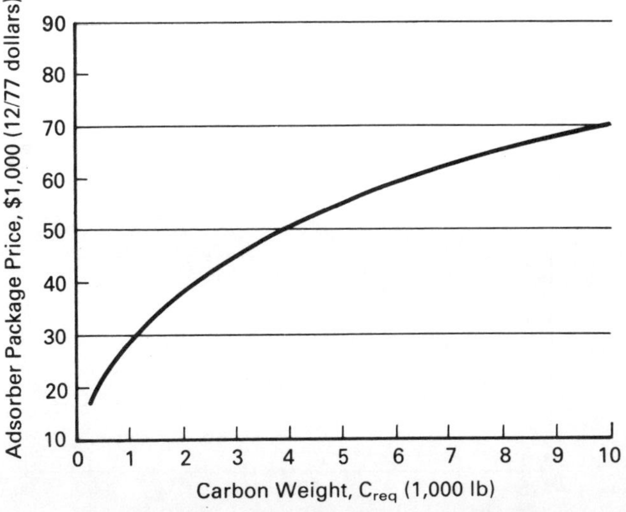

Price includes carbon, beds, fan and motor, and instrumentation and controls.

Figure 5-5. **Prices for custom carbon adsorbers, less carbon.(5)**

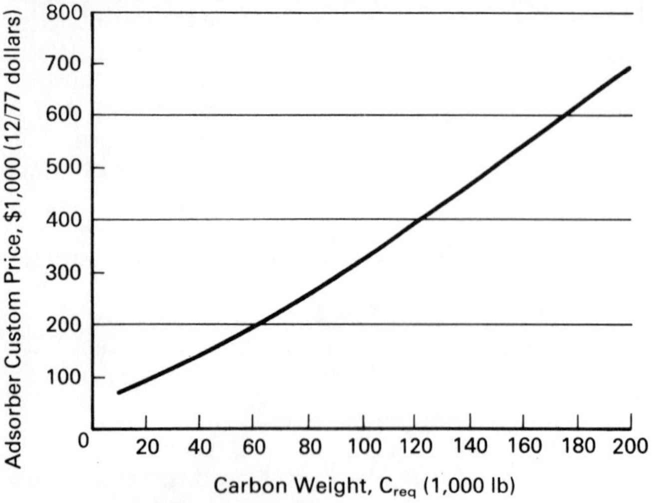

Price includes beds, instrumentation and controls, and a steam regenerator.

Figure 5-7. **Prices for absorber platform and ladders.(6)**

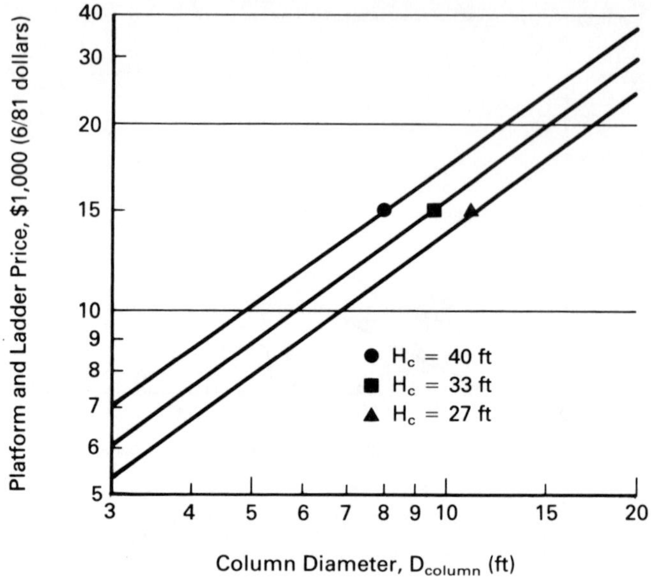

All auxiliary equipment (ductwork, fan, and stack) purchased costs and freight and taxes must be added to obtain the total purchased cost.

The cost curve for *absorbers* (Figure 5-6) does not include the cost of packing, platforms, and ladders. The cost of platform and ladders (Figure 5-7) and packing (Table 5-4) must be added to obtain the major equipment purchased cost. All auxiliary equipment (ductwork, fan, and stack) purchased costs, the cost of instrumentation and controls, and freight and taxes must be added to obtain the total purchased cost.

Figure 5-6. **Prices for absorber columns.(6)**

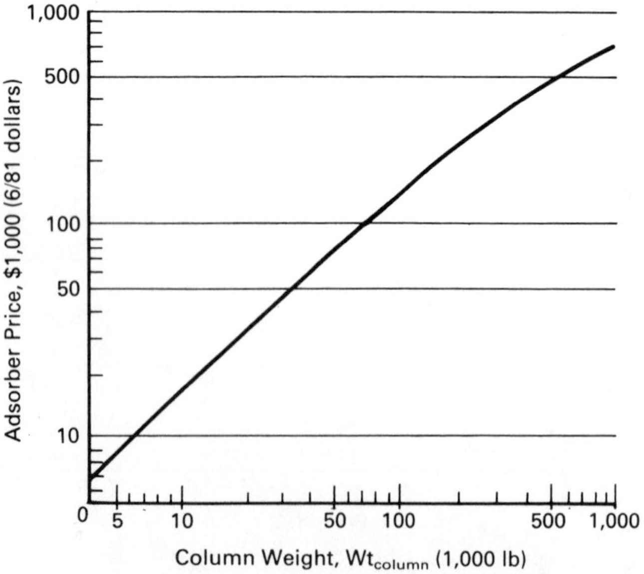

Price includes manholes, skirts, and painting.

Table 5-4. **Price of Packing for Absorber Systems (6)**

Packing Type and Material	Cost/Ft^3 (6/81 Dollars)			
Packing Diameter, inches	1	1.5	2	3
Pall Rings				
Carbon steel	24.3	16.5	15.1	—
Stainless steel	92.1	70.3	0.8	—
Polypropylene	21.9	14.8	13.8	—
Berl Saddles				
Stoneware	28.1	21.7	—	—
Porcelain	34.5	25.6	—	—
Intalox Saddles				
Polypropylene	21.9	—	13.6	7.0
Porcelain	19.4	14.8	13.3	12.2
Stoneware	18.2	13.3	12.2	11.0
Packing Rings				
Carbon steel	30.3	19.8	17.0	13.9
Porcelain	13.2	10.6	9.7	8.1
Stainless steel	109.0	82.6	22.9	—

The cost curve for *condensers* (Figure 5-8) yields the total capital cost for cold water condenser systems. For systems needing refrigerant (ethylene glycol), the applicable cost from Figure 5-9 must be added to the cost obtained from Figure 5-8. Since a total capital cost is determined, no additional cost estimates are necessary; therefore, proceed to Section 5.3 to calculate annualized operating costs.

The cost curve for a negative pressure *fabric filter* (Figure 5-10) does not include the the cost of bags (Table 5-5), which depend upon type of fabric used. This cost must be added to obtain the major equipment purchased cost. All auxiliary equipment (ductwork, fan, and stack) purchased costs, the cost

Figure 5-8. Total capital costs for cold water condenser systems.(7)

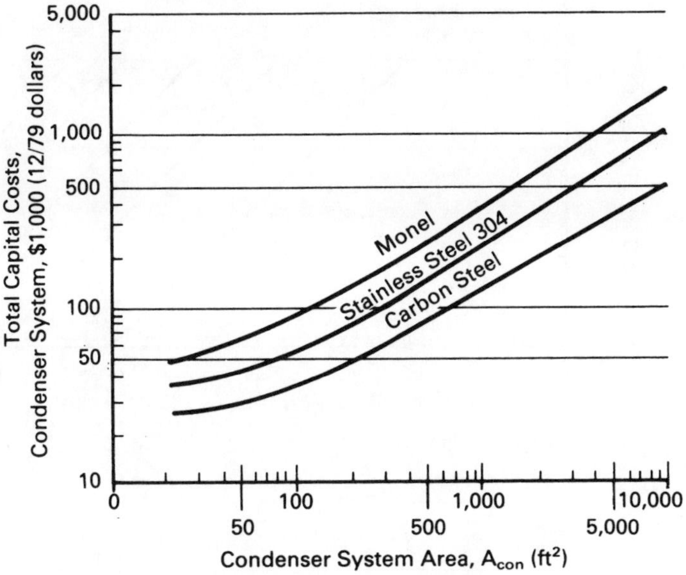

Figure 5-9. Additional capital cost for refrigerant condenser systems.(7)

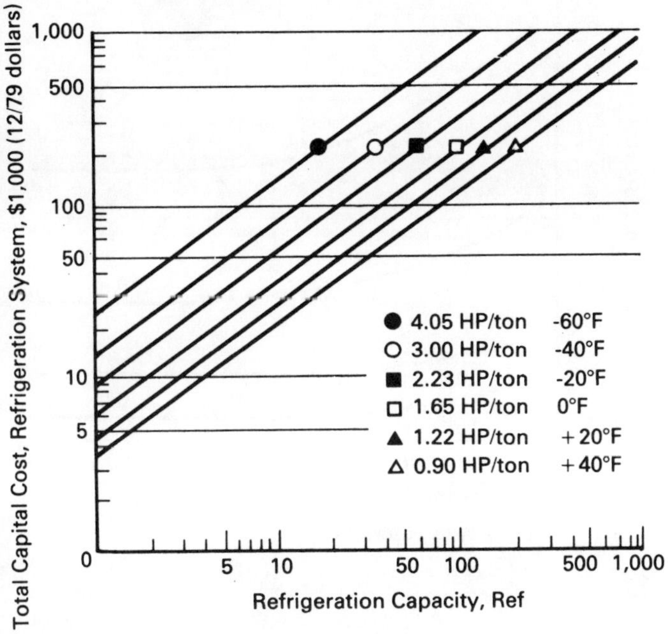

Figure 5-10. Prices for negative pressure, insulated fabric filter systems, less bags.(2)

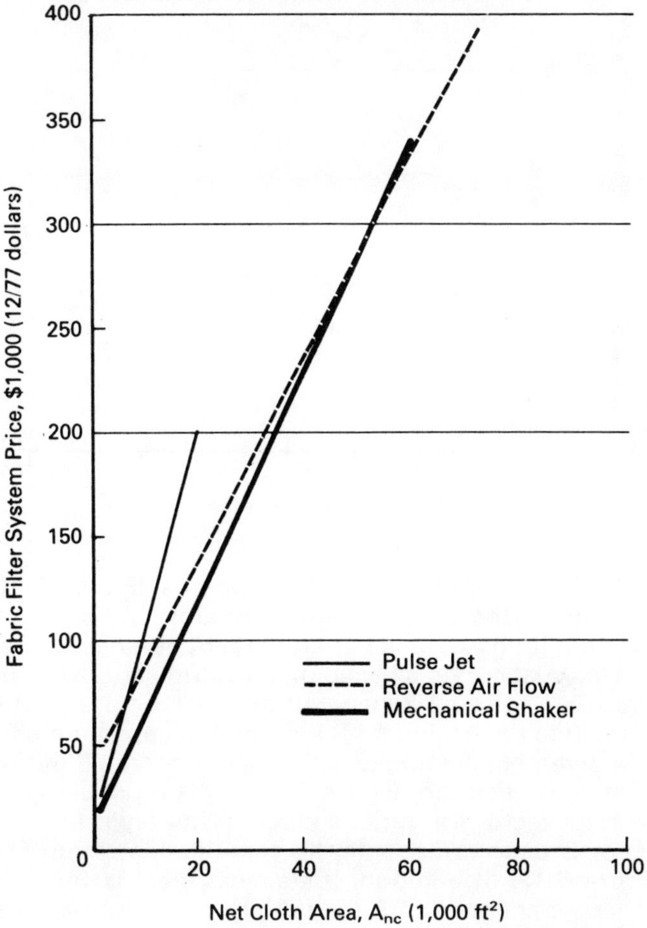

of instrumentation and controls, and freight and taxes must be added to obtain the total purchased cost.

The cost curve presented in Figure 5-11 provides the major equipment purchased cost for an insulated *electrostatic precipitator*. All auxiliary equipment (ductwork, fan, and stack) purchased cost, the cost of instrumentation and controls, and freight and taxes must be added to obtain total purchased cost.

Table 5-5. Bag Prices (2)

Class	Type	12/77 Dollars/Gross Ft^2						
		Dacron	Orlon	Nylon	Nomex	Glass	Polypropylene	Cotton
Standard	Mechanical Shaker <20,000 ft^2	0.40	0.65	0.75	1.15	0.50	0.65	0.45
Standard	Mechanical Shaker >20,000 ft^2	0.35	0.50	0.70	1.05	0.45	0.55	0.40
Standard	Pulse-jet[a]	0.60	0.95	—	1.30	—	0.70	—
Custom	Mechanical Shaker	0.25	0.35	0.45	0.65	0.30	0.35	0.40
Custom	Reverse Air	0.25	0.35	0.45	0.65	0.30	0.35	0.40

[a] For heavy felt, multiply source by 1.5.

Figure 5-11. Prices for insulated electrostatic precipitators.(2)

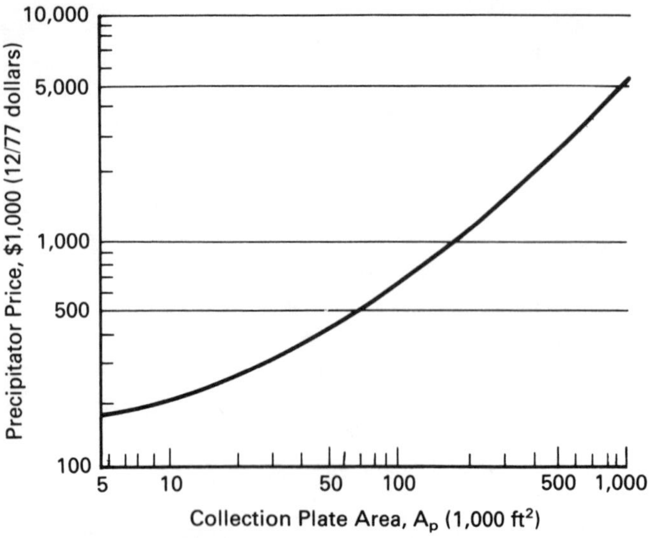

The cost curve for *venturi scrubbers* (Figure 5-12) includes the cost of instrumentation and controls, in addition to the major equipment purchased cost. This cost curve is based on a venturi scrubber constructed from 1/8-inch carbon steel. Figure 5-13 is used to determine if 1/8-inch steel is appropriate for a given application (use the higher curve). If thicker steel is required, Figure 5-14 yields a price adjustment factor for various steel thicknesses; this factor is used to escalate the cost obtained from Figure 5-12. In addition, if stainless steel is required (see Section 4.11.3.2) multiply the scrubber cost estimate by 2.3 for 304L stainless steel or by 3.2 for 316L stainless steel. Costs of all auxiliary equipment (ductwork, fan, and stack) and freight and taxes must be added to obtain the total purchased cost.

Figure 5-12. Prices for venturi scrubbers.(2)

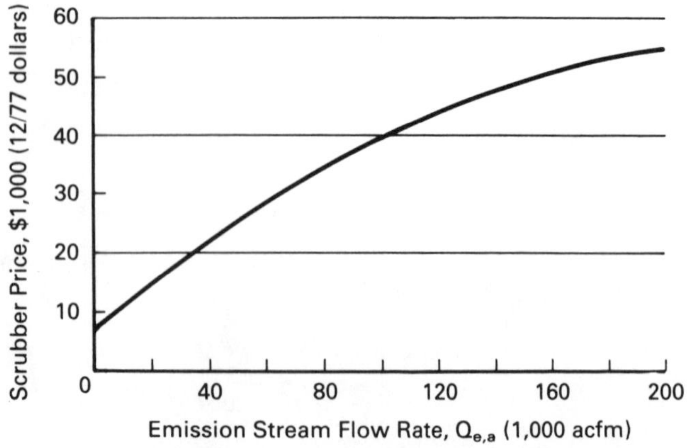

Price includes scrubber, elbows, separator, pumps, and instrumentation and controls. Price based on ⅛ in carbon steel.

Figure 5-13. Required steel thicknesses for venturi scrubbers.(2)

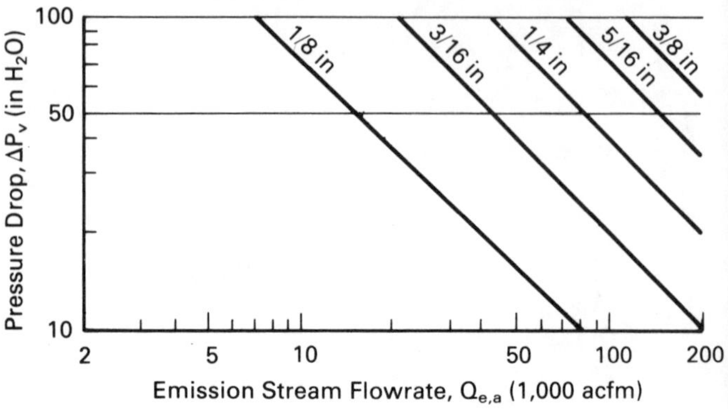

For use with Figure 5-12.

Figure 5-14. Price adjustment factors for venturi scrubbers.(2)

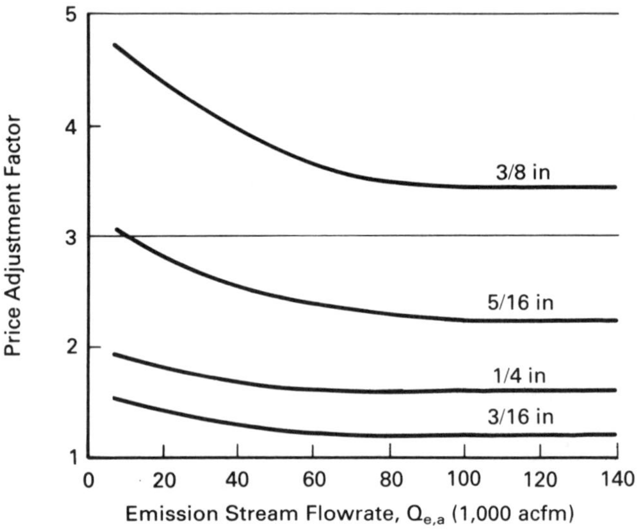

For use with Figure 5-12.

5.2.2 Estimation of Auxiliary Equipment Purchase Cost

The auxiliary equipment purchase cost is related to emission stream and equipment parameters. Table 5-6 presents the parameters that must be known for costing the auxiliary equipment. Figures 5-15 through 5-19 present the December 1977 costs for ductwork, fans, and stacks. The cost information presented in the figures are from available published data and must be escalated to reflect June 1985 dollars.

5.2.2.1 Ductwork Purchase Cost

The ductwork purchase cost is typically proportional to the ductwork weight, which is a function of: (1) the material of construction, (2) length, (3) diameter, and (4) thickness. Carbon steel ducts are normally used for noncorrosive flue gases at temperatures below 1,150°F. Stainless steel ducts are generally used with gas temperatures between 1,150°F to 1,500°F, or if the gas stream contains corrosive materials. Figures 5-15 and 5-16 present purchase costs for carbon steel and stainless steel ducts, respectively. It is assumed that the major portion of ductwork is utilized to transport the emission stream from the process to the control system; therefore, the flow rate of the emission stream at actual conditions ($Q_{e,a}$) is used to size the ductwork.

Without specific information, assume the following items to simplify the costing procedure:

(1) The ductwork is constructed with 3/16-inch thick plate.
(2) The duct length equals 100 feet.
(3) The duct diameter is calculated using a duct gas velocity of 2,000 ft/min. Therefore:

$$D_{duct} = 12\left(\frac{4}{\pi}\,\frac{Q_{e,a}}{U_{duct}}\right)^{\frac{1}{2}} = 0.3028\,(Q_{e,a})^{\frac{1}{2}} \qquad (5\text{-}1)$$

where:

D_{duct} = duct diameter, in
$Q_{e,a}$ = emission stream flow rate of actual conditions, acfm
U_{duct} = velocity of gas stream in duct, ft/min

Example Case

In the example case, since no specific data on the ductwork are available, use the above assumptions to cost the ductwork. The duct diameter is calculated according to item (3) above. The emission stream flow rate at actual conditions is approximately 16,500 acfm; therefore, the duct diameter equals 39 inches.

$$D_{duct} = 0.3028\,(16{,}500)^{\frac{1}{2}} = 39 \text{ in}$$

The length of the ductwork is assumed to be 100 feet of 3/16-inch thick plate. Since the emission stream treated contains no chlorine or sulfur compounds (i.e., it is a noncorrosive emission stream) and the gas temperature is 960°F, carbon steel ductwork is used. From Figure 5-15, the cost of the ductwork is estimated as follows:

\$52/ft x 100 ft x (336.2/226.2) = \$7,700

(Note: 12/77 dollars escalated to 6/85 dollars.)

Table 5-6. Identification of Design Parameters and Cost Curves for Auxiliary Equipment

Auxiliary Equipment	Design Parameters	Cost Curve Figure No.
Ductwork	Diameter	5-15
	Length	5-16
	Material of construction	
Fan[a]	Actual air flow rate	5-17
	Pressure drop	
	Gas stream velocity	
Stack	Diameter	5-18
	Length	5-19
	Material of construction	

[a] Assumed to be located downstream of the control system and ductwork.

Figure 5-15. Carbon steel straight duct fabrication price at various thicknesses.(2)

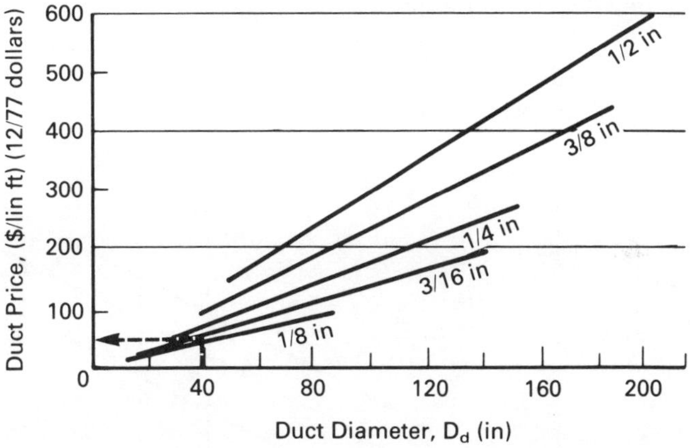

Figure 5-16. Stainless steel straight duct fabrication price at various thicknesses.(2)

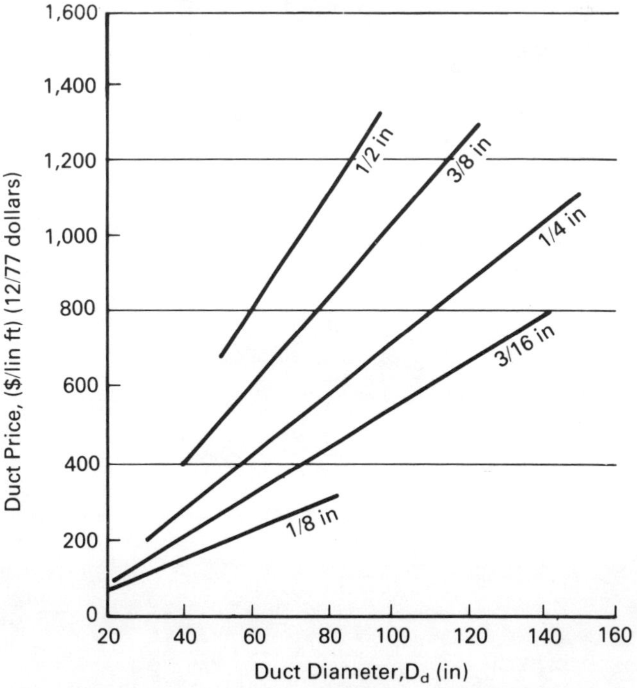

5.2.2.2 Fan Purchase Cost

The fan purchase cost (Figure 5-17) is a function of the flow rate moved by the fan and the pressure drop (ΔP) across the control system. The fan is assumed to be located downstream of the final control device in the control system. Therefore, the fan capacity must be based on the final control device's exit gas flow rate at actual conditions ($Q_{fg,a}$). Control system pressure drop (ΔP) is the total of the pressure drops across the various control system equipment, including the stack and ductwork. Table 5-7 presents conservative pressure drops across specific control system components which can be used if specific data are not available.

Using the actual flow rate and total ΔP parameters, obtain the fan purchased cost from Figure 5-17. Fans are categorized into Classes I to IV according to control system pressure drop. Guidelines are presented in Figure 5-17 to determine which class of fan to use. There is some overlap between the classes. The lower class fan is generally selected due to cost savings. To estimate the cost of a motor for the fan, multiply the fan cost by 15 percent. (Note: The fan and motor costs are included in the cost curves for thermal incinerators and packaged carbon adsorbers.)

Figure 5-17. Fan prices.(2)

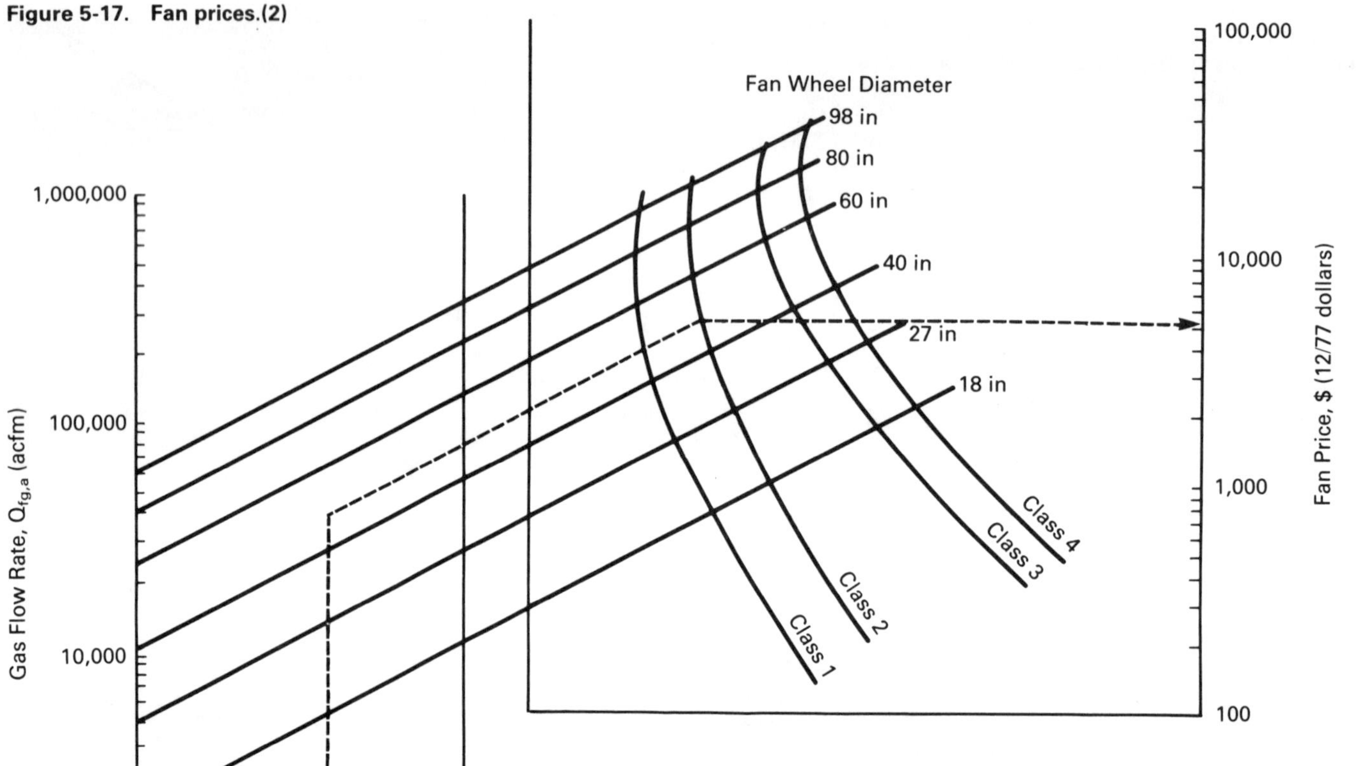

Fan Class	Performance Range[a] Single Width
I	5 in @ 2,300 fpm to 2-½ in @ 3,200 fpm
II	8-½ in @ 3,000 fpm to 4-¼ in @ 4,175 fpm
III	13-½ in @ 3,780 fpm to 6-¾ in @ 5,260 fpm
IV	Above Class III specifications

[a]Performance Range designations are indicated by static pressure (inches of water) at fan outlet velocity (feet per minute).

Notes:
Fan price is lower at higher pressure drop because smaller fan wheel is used at higher rpm.
For high temperature environment add 3% (250°F to 600°F).
For stainless steel construction multiply price by 2.5.

Figure 5-18. Carbon steel stack fabrication price for 1/4 in plate.(2)

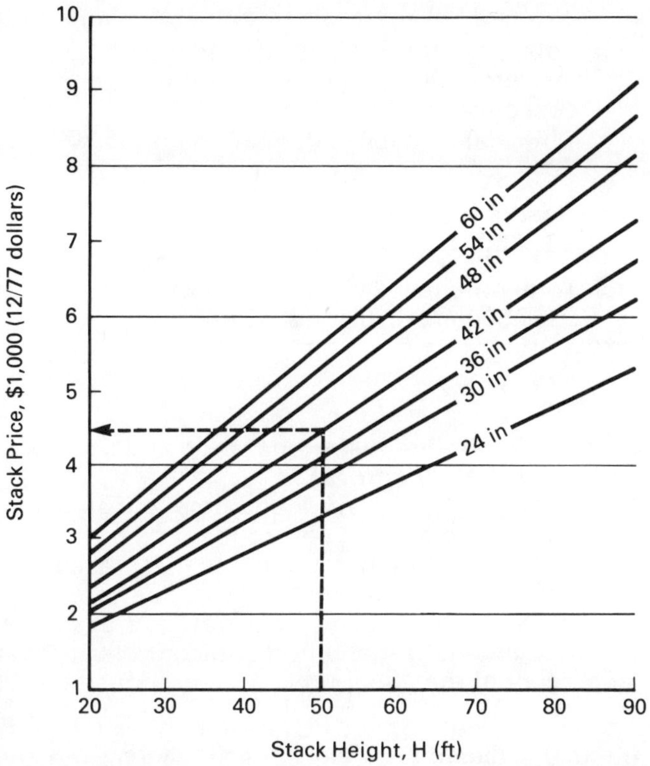

Figure 5-19. Carbon steel stack fabrication price for 5/16 in and 3/8 in plate.(2)

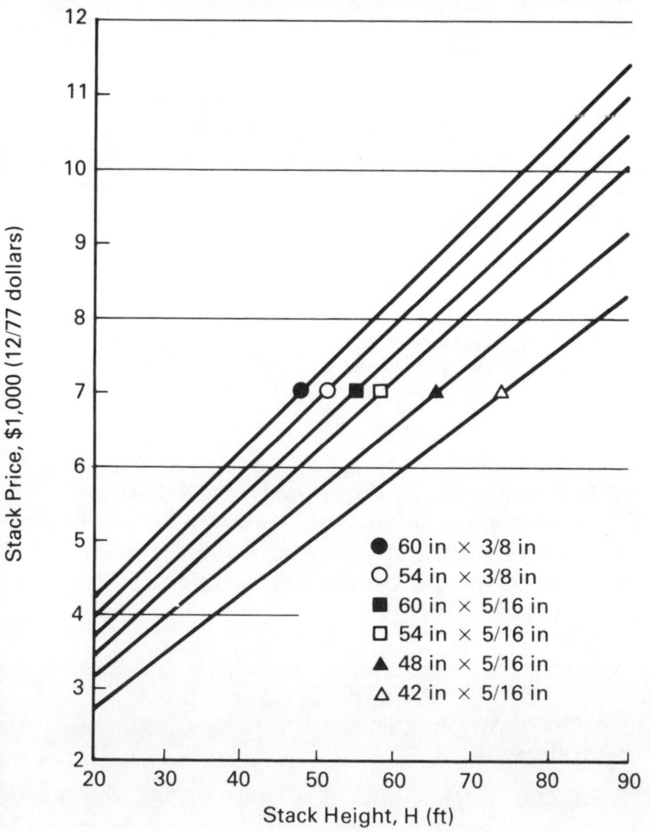

Table 5-7. Assumed Pressure Drops Across Various Components

System Component	Pressure Drop (in H_2O)
Stack	0.5
Ductwork	0.5
Thermal Incinerator	4
Heat Exchanger	2
Catalytic Incinerator	6
Absorber	Variable[a]
Carbon Adsorber	6
Condenser	3
Fabric Filter	6
Electrostatic Precipitator	0.5
Venturi Scrubber	ΔP_v

[a]Use Equation 4.7-21 (Section 4.7) to determine pressure drop.

Example Case

For the example case, the fan and motor costs (Figure 5-17) are included in the thermal incinerator cost curve; however, these costs can be calculated separately. The total pressure drop across the control system is 7 in H_2O (obtained from summing the values from Table 5-7 for the incinerator, heat exchanger, ductwork, and stack). The flow rate exiting the heat exchanger is approximately 40,000 acfm ($Q_{fg,a}$). The pressure drop from the guidelines on Figure 5-17 indicates that a Class II fan (the lower class fan) is appropriate. The estimated fan and motor costs are as follows:

(a) Fan cost—
$5,000 x (336.2/226.2) = $7,400
(Note: 12/77 dollars escalated to reflect 6/85 dollars.)

(b) Motor cost—
$7,400 x 0.15 = $1,100

5.2.2.3 Stack Purchase Cost

The stack purchase cost is a function of: (1) the material of construction, (2) height, (3) diameter, and (4) stack thickness. In addition, minimum stack exit velocities should be at least 1.5 times the expected wind velocity; or for instance, in the case of 30 mph winds, the minimum exit velocity should be at least 4,000 ft/min.[2] For purposes of this handbook, the stack is designed and costed with respect to the final control device's exit gas flow rate at actual conditions ($Q_{fg,a}$). Figures 5-18 and 5-19 present purchased costs for unlined, carbon steel stacks.

Without specific information, assume the following items to simplify the costing procedure:

(1) The stack is constructed with 1/4-inch thick carbon steel plate.
(2) The stack height equals 50 feet.

(3) The stack diameter is calculated using a stack exit velocity of 4000 ft/min. Therefore:

$$D_{stack} = 12\left(\frac{4}{\pi} \times \frac{Q_{fg,a}}{U_{stack}}\right)^{\frac{1}{2}} = 0.2141\,(Q_{fg,a})^{\frac{1}{2}} \quad (5\text{-}2)$$

where:

D_{stack} = stack diameter, in
$Q_{fg,a}$ = flue gas flow rate at actual conditions, acfm
U_{stack} = velocity of gas stream in stack, ft/min

Example Case

In the example case, since no specific data for the stack are available, use the above assumptions to cost the stack. The stack diameter is calculated according to item (3) above. The actual gas flow rate exiting the heat exchanger (flue gas flow rate) equals approximately 40,000 acfm. Therefore, the stack diameter is calculated as follows:

$D_{stack} = 0.2141\,(40{,}000)^{\frac{1}{2}} = 43$ in

With the stack diameter known, use the appropriate curve in Figure 5-18 (use the closest curve: 42 inches) to estimate the stack cost as follows:

$4,500 x (336.2/226.2) = $6,700

(Note: 12/77 dollars escalated to reflect 6/85 dollars.)

5.2.3 Estimation of the Total Purchased Equipment Cost

The total purchased equipment cost equals the sum of the major equipment purchased cost and the auxiliary equipment purchased cost. (Note: The major equipment purchased cost curves for thermal incinerators, carbon adsorbers, and venturi scrubbers also include the cost of instrumentation and controls; therefore, this cost must be subtracted to estimate the total purchased equipment cost for these control devices.) Calculate the total purchased equipment cost for thermal incinerators, carbon adsorbers, and venturi scrubbers as follows: (1) multiply the summation of the major equipment purchased cost and the auxiliary equipment purchased cost by a factor of 0.091 to obtain the cost of instrumentation and controls (the cost of instrumentation and costs is estimated to equal 10 percent of the total purchased equipment); and (2) subtract this cost from the summation of the major equipment purchased cost and the auxiliary equipment purchased cost.

Example Case

For the example case, the total purchased equipment cost is estimated as follows:

(a) Total purchased equipment cost plus cost of instrumentation and controls (included in cost curve)—
$145,700 + $104,400 + $7,700 + $6,700 = $264,500

(b) Cost of instrumentation and controls—
$264,500 x 0.091 = $24,100

(c) Total purchased equipment cost—
$264,500 - $24,100 = $240,400

5.2.4 Estimation of Instrumentation and Controls Plus Freight and Taxes

For the majority of control equipment, instrumentation costs are a small part of the total purchased cost. Instrumentation requirements for a control system depend upon control and safety requirements. When no specific cost data are available, estimate the instrumentation and controls costs at 10 percent of the total purchased equipment cost.

The cost of equipment freight and taxes depends upon the location of the control system and the location of the supplier. Without specific data, estimate the cost of freight and taxes at 8 percent of the total purchased equipment cost.

Example Case

For the example case, the cost of instrumentation and controls and the cost of freight and taxes are as follows:

(a) Instrumentation and controls—
$240,400 x 0.10 = $24,000

(b) Freight and taxes—
$240,400 x 0.08 = $19,200

5.2.5 Estimation of Total Purchased Cost

The summation of the total purchased equipment cost, the cost of instrumentation and controls, and the cost of freight and taxes equals the total purchased cost.

Example Case

The total purchased cost for the example case is as follows:

$240,400 + $24,000 + $19,200 = $283,600

5.2.6 Calculation of Total Capital Costs

The sum of the total purchased equipment cost, other direct costs, indirect costs, and contingency costs represents the total capital cost. Obtain the total capital cost for the control system by multiplying the total purchased cost by the appropriate factor listed in Table 5-8. This factor accounts for the other direct costs, the indirect costs, and the contingency costs. For control systems employing multiple control devices, use the largest applicable factor. Retrofit applications will likely be more expensive (see footnote e, Table 5-8). Each component that comprises the total capital cost is listed in Table 5-8 to allow insertion of specific cost data if they are known.

> **Example Case**
>
> The example case consists of a control system using one control device: a thermal incinerator. Therefore, by using the factor of 1.63 for thermal incinerators from Table 5-8, the total capital cost for the example case is as follows:
>
> $283,600 x 1.63 = $462,300

5.3 Annualized Operating Costs

The annualized cost of a control system can be divided into direct operating costs, indirect operating costs, and credits. In this handbook, the inflation effect on costs is not considered, annualized costs are assumed to be constant in real dollars, and the total annualized cost is estimated on a before-tax basis. The direct operating costs consist of utilities, operating labor charges, maintenance charges, and replacement parts and labor charges. Utilities (i.e., fuel, electricity, water, steam, and materials required for the control system) are annual costs that vary depending upon the control system size and operating time. They are calculated using gas stream characteristics and control equipment capacity data. Operating labor costs consist of operator labor and supervision, while maintenance costs consist of maintenance labor and materials. The direct operating costs are established by estimating annual quantities of utilities consumed and operator and maintenance labor used and by applying unit costs to these quantities. The annual quantities of utilities and labor requirements are assumed to be proportional to the annual operating hours for the control system. Operating labor supervision and maintenance materials are taken as percentages of the operator and maintenance labor costs. Costs of replacement parts are estimated as applicable, and the cost of replacement labor is assumed to equal the cost of replacement parts.

The indirect operating costs include overhead costs, property tax, insurance, administration costs, and the capital recovery costs. Overhead costs are estimated as a percent of operating labor

Table 5-8. Capital Cost Elements and Factors[a] (2)

	Control Technique						
Cost Elements	ESP	Venturi Scrubbers	Fabric Filters	Thermal & Catalytic Incinerators	Adsorbers	Absorbers	Condensers
DIRECT COSTS							
Purchased Equipment Cost[b]	1.00	1.00	1.00	1.00	1.00	1.00	1.00
Other Direct Costs:							
Foundation and supports	0.04	0.06	0.04	0.08	0.08	0.12	0.08
Erection and handling	0.50	0.40	0.50	0.14	0.14	0.40	0.14
Electrical	0.08	0.01	0.08	0.04	0.04	0.01	0.08
Piping	0.01	0.05	0.01	0.02	0.02	0.30	0.02
Insulation	0.02	0.03	0.07	0.01	0.01	0.01	0.10
Painting	0.02	0.01	0.02	0.01	0.01	0.01	0.01
Total Direct Cost	1.67	1.56	1.72	1.30	1.30	1.85	1.43
INDIRECT COSTS							
Engineering and supervision	0.20	0.10	0.10	0.10	0.10	0.10	0.10
Construction and field expenses	0.20	0.10	0.20	0.05	0.05	0.10	0.05
Construction fee	0.10	0.10	0.10	0.10	0.10	0.10	0.10
Start up	0.01	0.01	0.01	0.02	0.02	0.01	0.02
Performance test	0.01	0.01	0.01	0.01	0.01	0.01	0.01
Model study	0.02	—	—	—	—	—	—
Total Indirect Cost	0.54	0.32	0.42	0.28	0.28	0.32	0.28
CONTINGENCY[c]	0.07	0.06	0.07	0.05	0.05	0.07	0.05
TOTAL[d]	2.27	1.94	2.21	1.63	1.63	2.24	1.76

[a] As fractions of total purchased equipment cost. They must be applied to the total purchased equipment cost.

[b] Total of purchased costs of major equipment and auxiliary equipment and others, which include instrumentation and controls at 10%, taxes and freight at 8% of the equipment purchase cost.

[c] Contingency costs are estimated to equal 3% of the total direct and indirect costs.

[d] For retrofit applications, multiply the total by 1.25.

and supervision costs plus maintenance labor costs. Property tax, insurance, and administrative costs are estimated as a percent of the total capital cost. The capital recovery cost is estimated as the product of the capital recovery factor times the total capital cost. The factor for capital recovery costs (the total of annual depreciation and interest on capital) is determined from the expected life of the control device and the interest rate at which the capital is borrowed. The expected life of a given control device depends on the type of control application, maintenance service, and operating duty. For costing purposes, preestablished expected life values are used.

Some control techniques recover the HAP's from a given emission stream as a salable product. Therefore, any cost credits associated with the recovered material must be deducted from the total annualized cost to obtain the net annualized cost for the system. The amount, purity, and commercial value of the recovered material determine the magnitude of credits.

5.3.1 Direct Operating Costs

Table 5-9 presents June 1985 unit costs for utilities, operator labor, and maintenance labor as well as cost factors for other direct operating cost elements. The procedure used to estimate direct operating costs (including utilities, direct labor, maintenance, and replacement costs) and indirect operating costs (including overhead, property tax, insurance, administration, and capital recovery cost) was taken directly from reference 2. These unit costs and cost factors are applied to estimated quantities of utilities consumed, labor expended, and parts used to obtain total direct operating costs.

Table 5-9. Unit Costs to Calculate Annualized Cost

Cost Elements	Unit Costs/Factor
DIRECT OPERATING COSTS	
1. Utilities:[a]	
a. Natural gas	$ 0.00425 per ft^3 (11)
b. Fuel oil	$ 1.025 per gal (12)
c. Water	$ 0.0003 per gal (2)
d. Steam	$ 0.00504 per lb (2)
e. Electricity	$ 0.059 per kWh (13)
f. Solvent	As applicable
2. Operating Labor:	
a. Operator Labor	$ 11.53 per hour (14)
b. Supervision	15% of Operator Labor
3. Maintenance:	
a. Labor	$ 11.53 per hour (14)
b. Materials[b]	100% of Maintenance Labor
4. Replacement:	
a. Parts	As applicable (see Table 5-10)
b. Labor	100% of Replacement Parts
INDIRECT OPERATING COSTS	
1. Overhead	80% of 2a + 2b + 3a + 4a
2. Property Tax	1% of Total Capital Cost
3. Insurance	1% of Total Capital Cost
4. Administration	2% of Total Capital Cost
5. Capital Recovery	(CRF[c]) x Total Capital Cost
CREDITS	As applicable

[a] Refer to Tables 5-10 and 5-11 to estimate utility costs for each HAP control technique.

[b] Maintenance materials include operating supplies (e.g., lubrication, paper).

[c] CRF = capital recovery factor.
For an average interest rate of 10%, the CRF for specific control devices are listed below.
ESP and fabric filter: CRF = 0.117 (based on 20-year life span).
Venturi scrubber, thermal and catalytic incinerators, adsorber, absorber, and condenser: CRF = 0.163 (based on 10-year lifespan).

If a given control system contains two or more control devices, the direct operating costs must be calculated for each device and summed. The capital recovery cost for a multiple control device system should be calculated using a weighted average capital cost factor.

Unless specified, use 8,600 hours per year, 8 hours per shift, and 24 hours per day, as necessary, to estimate the annual costs for utilities consumed, operator labor, and maintenance labor.

5.3.1.1 Determine Utility Requirements

The utility requirements for a control system are obtained from each component's design calculations. Use the costing information in Table 5-9, Table 5-10, and Table 5-11 to estimate the total utility costs. A procedure to estimate fan electricity costs is provided below, since these costs are applicable to all control techniques. The fan horsepower requirements are calculated as follows:

$$\text{Fan horsepower, HP} = \frac{0.000157}{\eta} \times Q_{fg,a} \times \Delta P \quad (5\text{-}3)$$

where:
HP = fan horsepower requirement, hp
ΔP = pressure drop across the control system, in H_2O
η = fan efficiency (usually 60-70%)

Assuming a 65 percent fan efficiency and a 10 percent additional capacity requirement for miscellaneous purposes, and using the conversion factor of 0.746 kilowatt hour per horsepower-hour, estimate the fan electricity requirement as follows:

$$\text{FER} = 2.0 \times 10^{-4} \, (Q_{fg,a}) \, (\Delta P) \, (\text{HRS}) \quad (5\text{-}4)$$

where:
FER = fan electricity requirement, kWh
HRS = hours of operation per year

5.3.1.2 Determine Remaining Direct Operating Costs

The remaining direct operating costs include replacement parts and labor, operating labor (i.e., the summation of operator labor and supervision labor), and maintenance (i.e., the summation of

Example Case

For the thermal incinerator example case, since no specific information is available, the control system is assumed to operate 8,600 hours per year. According to the design calculations in Section 4.2, and Table 5-10, the only utility requirement for the thermal incinerator, in addition to fan electricity, is 330 scfm of natural gas. Using Equation 5-4, the electricity requirement for a 40,000 acfm fan at 7 in H_2O pressure drop is as follows:

2.0×10^{-4} (40,000 acfm) (7 in) (8,600 hr) = 481,600 kWh

Applying the unit costs from Table 5-9 and the equation for fuel requirements from Table 5-11, the utility costs are:

(a) Natural gas cost—
60 (330 scfm)(8,600 hr)($0.00425/ft^3) = $723,700

(b) Electricity cost—
481,600 kWh ($0.059/kWh) = $28,400

maintenance labor and materials). Tables 5-9 and 5-10 provide the necessary information to calculate the cost of replacement parts and labor. Table 5-12 presents available data on estimated labor requirements for various control devices. The labor requirements presented as "hours per shift" must be converted to annual requirements. Total annual operator and maintenance labor costs are obtained by multiplying the estimated annual labor requirements with the applicable unit costs from Table 5-9. These costs must be determined for each control device in the control system. Operating labor supervision and maintenance materials are estimated as a percentage of operator labor and maintenance labor, respectively. Again, these costs must be determined for each control device if a multiple control device system is used.

Table 5-10. Utility/Replacement Operating Costs for HAP Control Techniques[a]

HAP Control Device	Utilities/Replacement Parts
Thermal Incinerator	Natural gas or fuel oil[b] Electricity (fan)
Catalytic Incinerator	Catalyst cost[c] (V_{cat}) Natural gas or fuel oil[b] Electricity (fan)
Carbon Adsorber Systems	Carbon[d] (C_{req}) Steam[b] Cooling water[b] Electricity (fan)
Absorber Systems	Absorbent[b] (water or solvent) Electricity (fan)
Condenser System	Refrigerant[e] (Ref) Electricity (fan)
Fabric Filter Systems	Bags[f] (A_{tc}) Electricity[b] (fan + control device)
Electrostatic Precipitators	Electricity[b] (fan + control device)
Venturi Scrubbers	Water[b] Electricity (fan)

[a] Refer to Table 5-9 for utility unit costs, Tables 5-3 and 5-5 for replacement part unit costs, and Table 5-2 for FE cost indices.
[b] See Table 5-11.
[c] Annualized replacement catalyst costs are calculated as follows:

$$\text{Annualized cost} = \frac{V_{cat}\,(ft^3) \times \$/ft^3}{3 \text{ years}} \text{ (Current FE/Base FE)}$$

[d] Annualized replacement carbon costs are calculated as follows:

$$\text{Annualized cost} = \frac{C_{req}\,(lb) \times \$/lb}{5 \text{ years}} \text{ (Current FE/Base FE)}$$

[e] Refrigerant replacement is due to system leaks, however, the loss rate of refrigerant is very low and varies for every unit. Therefore, assume that the cost of refrigerant replacement is negligible.
[f] Annualized replacement bag costs are calculated as follows:

$$\text{Annualized cost} = \frac{A_{tc}\,(ft^2) \times \$/ft^2}{2 \text{ years}} \text{ (Current FE/Base FE)}$$

Example Case

Table 5-10 indicates that thermal incinerators do not require replacement parts or labor. From Tables 5-9 and 5-12, both the estimated operator and maintenance labor requirements and their unit costs for thermal incinerator systems are 0.5 hours/shift and $11.53/hr, respectively. Therefore, the annual operator and maintenance labor costs for the example are the same and each cost is estimated as follows:

8,600 hr/yr x 0.5 hr/shift x 11.53/hr ÷ 8 hr/shift = $6,200

Estimated other direct operating costs for the example case using the Table 5-9 factors are as follows:

Operating Labor Supervision = $6,200 x 0.15 = $900
Maintenance Materials = $6,200 x 1.00 = $6,200

Therefore, the total annual direct operating cost for the example case is estimated by summing the utility costs and the remaining direct operating costs:

$723,700 + $28,400 + $6,200 + $900 + $6,200 + $6,200 = $771,600

Table 5-11. Additional Utility Requirements (2)

Fuel Requirement for Incinerators, ft^3
(Note: The design sections for thermal and catalytic incinerators are developed under the assumption that natural gas is used as the supplementary fuel. Fuel oil could be used, however, the use of natural gas is normal industry practice. If fuel oil is used, the equation below can be used by replacing Q_f with the fuel oil flow rate in units of gallons per minute. The product of the equation then equals gallons of fuel oil.)
Fuel Requirement = 60 (Q_f) x HRS
where:
Q_f = supplementary fuel required, scfm
HRS = annual operating hours, hr
(Note: Use 8,600 hours unless otherwise specified.)

Steam Requirement for Carbon Adsorber, lb
(Note: Assume 4 lb of steam required for each lb of recovered product.)
Steam Requirement = 4 (Q_{rec}) x HRS
where:
Q_{rec} = quantity of HAP recovered, lb/hr
HRS = annual operating hours, hr
(Note: Use 8,600 hours unless otherwise specified.)

Cooling Water Requirement for Carbon Adsorber, gal
(Note: Assume 12 gal of cooling water required per 100 lbs steam.)
Water Requirement = 0.48 (Q_{rec}) x HRS
where:
Q_{rec} = quantity of HAP recovered, lb/hr
HRS = annual operating hours, hr
(Note: Use 8,600 hours unless otherwise specified.)

Absorbent Requirement for Absorbers, gal
(Note: Assume no recycle of absorbing fluid [water or solvent].)
Absorbent Requirement = 60 (L_{gal}) x HRS
where:
L_{gal} = absorbing fluid flow rate, gal/min
HRS = annual operating hours, hr
(Note: Use 8,600 hours unless otherwise specified.)

Water Requirement for Venturi Scrubbers, gal
(Note: Assume 0.01 gal H_2O are required per acf of emission stream.)
Water Requirement = 0.6 ($Q_{e,a}$) x HRS
where:
$Q_{e,a}$ = emission stream flow rate into scrubber, acfm
HRS = annual operating hours, hr
(Note: Use 8,600 hours unless otherwise specified.)

Baghouse Electricity Requirement, kWh
(Note: Assume 0.0002 kW are required per ft^2 of gross cloth area.)
Baghouse Electricity Requirement = 0.0002 (A_{tc}) x HRS
where:
A_{tc} = gross cloth area required, ft^2
HRS = annual operating hours, hr
(Note: Use 8,600 unless otherwise specified.)

ESP Electricity Requirement, kWh
(Note: Assume 0.0015 kW are required per ft^2 of collection area.)
ESP Electricity Requirement = 0.0015 (A_p) x HRS
where:
A_p = collection plate area, ft^2
HRS = annual operating hours, hr
(Note: Use 8,600 unless otherwise specified.)

Table 5-12. Estimated Labor Hours per Shift and Average Equipment Life (2)

	Labor Requirements (hr/shift)		
Control Device	Operator Labor	Maintenance Labor	Average Equipment Life (yr)
Electrostatic Precipitator	0.5 - 2	0.5 - 1	20
Fabric Filter	2 - 4	1 - 2	20
Venturi Scrubber	2 - 8	1 - 2	10
Incinerator	0.5	0.5	10
Adsorber	0.5	0.5	10
Absorber	0.5	0.5	10
Condenser	0.5	0.5	10

5.3.2 Indirect Operating Costs

The indirect operating costs include overhead costs, property tax, insurance, administration, and the capital recovery cost. Estimate the overhead costs as 80 percent of the direct labor cost (the summation of operating labor and supervision labor costs) and the maintenance labor cost. The property tax estimate is calculated as 1 percent of the total capital cost, insurance as 1 percent of the total capital cost, and administration as 2 percent of the total capital cost. Estimate the capital recovery cost portion of the fixed capital charges by multiplying the total capital cost by a capital recovery factor. The capital recovery factor (CRF) is calculated as follows:

$$CRF = [i(1 + i)^n] / [(1 + i)^n - 1] \tag{5-5}$$

where:
i = interest rate on borrowed capital, decimal
n = control device life, years

For the purpose of this handbook, an interest rate of 10 percent is used. Table 5-12 contains data on expected control device life (n). Calculated capital recovery factors at 10 percent interest rate are 0.163, and 0.117 for 10- and 20-year control device lifetimes, respectively. If more than one control device is used by the control system, use a weighted average capital recovery factor. A weighted average capital recovery factor (CRFw) is determined as follows:

$$CRF_1 = CRF_1 [PC_1/(PC_1 + PC_2)] + CRF_2 [PC_2/(PC_1 + PC_2)] \tag{5-6}$$

where:
CRF_1 = the capital recovery factor for control device #1
CRF_2 = the capital recovery factor for control device #2
PC_1 = the purchased equipment cost for control device #1
PC_2 = the purchased equipment cost for control device #2

Example Case

As estimated in Section 5.2.5, the total capital cost for the thermal incinerator control system example is \$462,300. Section 5.3.1.2 estimated that the direct and maintenance labor cost is \$13,300 (i.e., \$6,200 + \$900 + \$6,200). Therefore, the indirect operating costs are estimated as follows:

(a) Overhead costs—
\$13,300 x 0.80 = \$10,600

(b) Property tax—
\$462,300 x 0.01 = \$4,600

(c) Insurance—
\$462,300 x 0.01 = \$4,600

(d) Administration charges—
\$462,300 x 0.02 = \$9,200

(e) Capital recovery—
The capital recovery factor is calculated using Equation 5-5: obtain the expected equipment lifetime for an incinerator ffrom Table 5-12 (10 years), and assume an interest rate of 10 percent.

$$[0.1\ (1 + 0.1)^{10}]\ /\ [(1 + 0.1)^{10} - 1] = 0.163$$

The capital recovery cost is then estimated as follows:

\$462,300 x 0.163 = \$75,400

(f) Total indirect operating costs—
\$10,600 + \$4,600 + \$4,600 + \$9,200 + \$75,400 = \$104,400

5.3.3 Credits

Credits for recovery of a salable product or energy must be included in determining the net annualized cost of the control system. The design calculations for the specific control devices include the quantity of recovered product. This information, along with product cost data from the inquirer/applicant, is used to calculate the credits on an annual basis.

Example Case

For the example case, there are no recovered products since a thermal incinerator destroys the organic vapors contained in the emission stream.

5.3.4 Net Annualized Costs

The direct and indirect operating costs less credits received equal the net annualized cost of the HAP control system.

Example Case

Total direct and indirect operating costs for the example case are \$771,600 and \$104,400, respectively. There are no recovery credits. Thus, the net annualized cost of the example HAP control system is as follows:

\$771,600 + \$104,400 - \$0 = \$876,000

5.4 References

1. U.S. EPA. *Identification, Assessment, and Control of Fugitive Particulate Emissions.* Draft final report. EPA Contract No. 68-02-3922. April 10, 1985.
2. U.S. EPA. *Capital and Operating Costs of Selected Air Pollution Control Systems.* EPA-450/5-80-002. December 1978.
3. U.S. EPA. *Organic Chemical Manufacturing Volume 4: Combustion Control Devices.* EPA-450/3-80-026. December 1980.
4. Vatavuk, W.M. and Neveril, R.B. "Estimate the Size and Cost of Incinerators." *Chemical Engineering.* July 12, 1982.
5. Vatavuk, W.M. and Neveril, R.B. "Costs of Carbon Adsorbers." *Chemical Engineering.* January 23, 1983.
6. Vatavuk, W.M. and Neveril, R.B. "Costs of Gas Adsorbers." *Chemical Engineering.* October 12, 1982.
7. U.S. EPA. Organic Chemical Manufacturing. *Volume 5: Adsorption, Condensation, and Absorption Devices.* EPA-450/3 80 077. December 1980.
8. Chemical Marketing Reporter. Schnell Publishing Company. New York, NY. Volume 228. September 16, 1985.
9. Telecon. Charlotte Clark (PES) to David Chalmers (Calgon Corp.). June 12, 1985.
10. Telecon. Gunseli Sagun Shareef (Radian Corp.) to Roy Uhlman (Engelhart Specialty Chemicals). September 30, 1985.
11. Phillip Gennarelli, Manager, American Gas Association (AGA). *Monthly Gas Utility Statistical Report.* AGA, 1515 Wilson Blvd., Arlington, Va. 22209. July 1985.
12. Bureau of Labor Statistics. August 1985. U.S. consumer price for fuel oil.
13. Monthly Energy Review. October 1982 electricity costs. DOE/EIA-003583/01. January 1983.
14. Survey of Current Business. November 1982. Primary Metal Industry Costs. December 1982.

Appendix A.1
Potential HAP's for Solvent Usage Operations

		Solvent Degreasing	Dry Cleaning	Graphic Arts[1]	Waste Solvent Reclaiming	SC[2]-Flatwood Paneling[3]	SC-Machinery[4]	SC-Appliances[5]	SC-Metal Furniture	SC-Auto/Truck[6]	SC-Fabrics	SC-Cans[7]	SC-Paper, Tapes, Labels	Magnetic Tape Coating	SC-Electrical Insulation	SC-Marine Vessels[8]	Vinyl & Acrylic Coatings[9]	SC-Wood Furniture	SC-Trans. Vehicles[10]	Machine Lubricants	Rubber Tire Manufacturing
Aliphatic Hydrocarbons	*Specific Compounds*																				
	Cyclohexane	•												•							
	Generic Compounds																				
	Naptha's mineral spirits	•	•	•	•	•	•	•	•	•	•	•	•	•	•	•	•	•	•		•
	Stoddard solvent			•																	
	Alicyclics		•				•														
Aromatic Hydrocarbons	*Specific Compounds*																				
	Benzene	•	•								•									•	
	Toluene	•		•		•	•	•	•	•		•	•	•		•		•	•		•
	Xylenes	•	•	•		•	•	•	•	•		•	•	•		•		•	•		•
	Napthalene[11]																				
	Generic Compounds																				
	Other aromatics[13]				•	•	•	•	•	•		•	•		•			•	•	•	
Halogenated[12] Hydrocarbons	*Specific Compounds*																				
	Chloromethane[11]																				
	Methylene chloride	•			•																
	Chloroform[11]																				
	Carbon tetrachloride	•	•	•																	
	1,1-dichloroethane										•										
	Trichloroethylene	•	•		•						•										
	1,1,1-trichloroethane	•																			
	Tetrachloroethylene	•	•																		
	Trichlorotrifluoroethane	•	•		•																
	Chlorobenzene	•									•										
	o,p-Dichlorobenzene	•									•										
	Generic Compounds																				
	Halogenated solvents	•			•																
Alcohols Glycols Ethers Epoxides Phenols	*Specific Compounds*																				
	Methanol			•		•				•											•
	Ethylene glycol			•		•		•		•		•	•						•		
	Propylene oxide											•									
	Cresols														•						
	Phenol			•																	
	Generic Compounds																				
	Alcohols	•		•		•	•	•	•	•		•	•	•				•	•		•
	Glycols								•											•	
	Cellusolves			•		•	•	•		•		•	•	•	•	•		•	•		
	Ethers	•		•			•	•													
	Phenols				•															•	
	Epoxides											•									

(continued on next page)

1. Category includes organic compounds associated with inks and solvents used in flexography, lithography, offset printing, and textile printing.
2. SC: surface coating.
3. Category includes coating of other flat stock.
4. Category includes coating of miscellaneous metal parts and coating of machinery and equipment.
5. Category includes all categories of appliances; large and small.
6. Category includes coating of automobiles and light duty trucks as well as automobile refinishing.
7. Category includes surface coating of coils, cans, containers, and closures.
8. Category includes coating of pleasure and commercial marine vessels and maintenance of vessels.
9. Category includes vinyl, acrylic, and nitrocellulose coatings.
10. Category includes surface coating of trucks, buses, railroad cars, and other transportation vehicles.
11. No information on specific categories using these compounds was located.
12. Appendix A.3, reference 1, provides a list of stabilizers that may be used in halogenated hydrocarbons.
13. Category includes polycyclic organic matter.

		Solvent Degreasing	Dry Cleaning	Graphic Arts[1]	Waste Solvent Reclaiming	SC[2]-Flatwood Paneling[3]	SC-Machinery[4]	SC-Appliances[5]	SC-Metal Furniture	SC-Auto/Truck[6]	SC-Fabrics	SC-Cans[7]	SC-Paper, Tapes, Labels	Magnetic Tape Coating	SC-Electrical Insulation	SC-Marine Vessels[8]	Vinyl & Acrylic Coatings[9]	SC-Wood Furniture	SC-Trans. Vehicles[10]	Machine Lubricants	Rubber Tire Manufacturing
Ketones Aldehydes	*Specific Compounds*																				
	Formaldehyde									●	●										
	Acetaldehyde[11]																				
	Furfural																●				
	Acetone	●		●		●	●		●	●			●			●		●	●		
	Acrolein (propenal)[11]																				
	Methyl ethyl ketone	●		●		●	●	●	●	●		●	●	●	●	●	●	●	●		
	Methyl isobutyl ketone			●		●	●		●	●		●	●	●		●		●	●		
	Cyclohexanone											●									
	Generic Compounds																				
	Aldehydes											●									
	Ketones			●		●	●	●	●	●		●	●					●	●		●
Esters Amides	*Specific Compounds*																				
	Ethyl acetate			●		●	●		●	●		●	●		●	●		●	●		
	Generic Compounds																				
	Esters			●		●	●	●	●	●		●	●	●				●	●	●	●
	Amides											●									
	Nitrosamines																			●	
Particulates	*Specific Compounds*																				
	Cadmium						●	●	●	●						●			●		
	Chromium						●	●	●	●						●			●		
	Lead						●	●	●	●						●			●		
	Zinc						●	●	●	●						●			●		
Acids Nitriles	*Specific Compounds*																				
	Nitrobenzene[11]																				
	Generic Compounds																				
	Organic acids									●											
	Nitriles				●																
	Nitrocompounds											●									
Heterocyclic Compounds	*Specific Compounds*																				
	Tetrahydrofuran											●		●							
	Furfural																●				
	Generic Compounds																				
	Pyrrolidones								●												
Miscellaneous Trade Solvents	So Cal I + II					●															
	Solvesso 100 + 150									●		●									
	Panasolve											●									
	Hi Sol 100											●									
	Tenneco T-125											●									

Appendix B.1
Gas Stream Parameters Calculations

At many plants, it is common that one pollution control system serves several emission sources. In such situations, the combined emission stream parameters must be calculated from mass and heat balances. Procedures for calculating the combined emission stream and single emission stream parameters listed below are provided in this appendix.

B.1.1 Flow Rate and Temperature
B.1.2 Moisture Content, SO_3 Content, and Dew Point
B.1.3 Particulate Matter Loading
B.1.4 Heat Content

B.1.1. Emission Stream Flow Rate and Temperature Calculations

Only gas volumes at standard conditions (70°F, 1 atm) can be added together. Thus, volumes of all gas streams must first be converted to volumes at standard conditions. This calculation is shown below. (Note: It is assumed that the emission streams are approximately at atmospheric conditions; therefore, pressure corrections are not necessary.)

$$Q_{e1,a} \times \frac{530}{460 + T_{e1}} = Q_{e1}$$

where:
$Q_{e1,a}$ = flow rate of gas stream #1 at actual conditions (acfm)
T_{e1} = temperature of gas stream #1 (°F)
Q_{e1} = flow rate of gas stream #1 at standard conditions (scfm)

This calculation is repeated for each emission stream which, when combined, will be served by the control system. The total gas stream volumetric flow rate at standard conditions (Q_e) is calculated by adding all gas streams, as follows:

$$Q_{e1} + Q_{e2} + \ldots = Q_e$$

where:
Q_e = flow rate of combined gas stream (scfm)

The temperature of the combined gas stream (T_e) must be calculated to convert this combined volumetric flow rate at standard conditions (Q_e) to actual conditions ($Q_{e,a}$).

The temperature of the combined gas stream (T_e) is determined by first calculating the enthalpy (sensible heat content) of each individual stream. The calculation procedures are shown below.

$$Q_{e1} \times \frac{0.018 \text{ Btu}}{\text{ft}^3 - {}^\circ\text{F}} \times (T_{e1} - 70) = H_{s1}$$

where:
T_{e1} = temperature of gas stream #1 (°F)
H_{s1} = sensible heat content of gas stream #1 (Btu/min)

This calculation is repeated for each emission stream. The total sensible heat is calculated as follows:

$$H_{s1} + H_{s2} + \ldots = H_s$$

where:
H_s = sensible heat of combined gas stream (Btu/min)

The combined gas stream temperature (T_e) is calculated as follows:

$$H_s \times \frac{\text{ft}^3 - {}^\circ\text{F}}{0.018 \text{ Btu}} \times \frac{1}{Q_e} = T_e$$

where:
T_e = temperature of combined gas stream (°F)

The actual combined gas stream volumetric flow rate at actual conditions ($Q_{e,a}$) is then determined as follows:

$$Q_e \times \frac{460 + T_e}{530} = Q_{e,a}$$

where:
$Q_{e,a}$ = flow rate of combined gas stream at actual conditions (acfm)

B.1.2 Moisture Content, SO_3 Content, and Dew Point Calculations

Moisture content is typically reported as a volume percent. The calculation procedures require that the volume percent moisture content of each stream be converted to a lb-mole basis, added together, and then divided by the total combined gas stream volumetric flow rate (Q_e) to obtain the moisture content of the combined gas stream. The moisture content is calculated below both on a volume percent and mass percent basis. The mass basis is to allow for the dew point calculation.

The moisture content is converted from a vol % basis to a lb-mole basis as follows:

$$M_{e1} \times \frac{1}{100\%} \times Q_{e1} \times \frac{\text{lb-mole}}{414 \text{ scf}} = M_{e1,lm}$$

where:
M_{e1} = moisture content of gas stream #1 (% vol)
$M_{e1,lm}$ = moisture content of gas stream #1 (lb-mole/min)

This calculation is repeated for each emission stream to be combined. The moisture content of the combined gas stream on a volume percent basis (M_e) is calculated by adding, as follows:

$$M_{e1,lm} + M_{e2,lm} + \ldots = M_{e,lm}$$

$$M_{e,1m} \times \frac{414 \text{ scf}}{\text{lb-mole}} \times \frac{1}{Q_e} \times 100\% = M_e$$

where
$M_{e,lm}$ = moisture content of combined gas stream (lb-mole/min)
M_e = moisture content of combined gas stream (% vol)

The moisture content of the combined stream must be reported on a mass basis ($M_{e,m}$) to determine the dew point. This is calculated as follows:

$$M_{e,lm} \times \frac{18 \text{ lb}}{\text{lb-mole}} = M_{e,m}$$

where:
$M_{e,m}$ = moisture content of combined gas stream (lb/min)

The amount of dry air in the combined gas stream (DA_e) is calculated as follows:

$$Q_e \times \frac{\text{lb-mole}}{414 \text{ scf}} \times \frac{29 \text{ lb}}{\text{lb-mole}} = DA_e$$

where:
DA_e = dry air content of combined gas stream (lb/min)

Calculate the psychrometric ratio as follows:

$M_{e,m}/(DA_e - M_{e,m})$ = psychrometric ratio (lb of water/lb dry air)

Knowing the psychrometric ratio and the gas stream temperature, the dew point temperature is selected from Table B.1-1.

The presence of sulfur trioxide (SO_3) in the gas stream increases the dew point of the stream. If the SO_3 component is ignored during the dew point determination, condensation may occur when not expected. In addition to the problems associated with the entrainment of liquid droplets in the gas stream, the SO_3 will combine with the water droplets to form sulfuric acid, which causes severe corrosion on metal surfaces and deterioration of many fabrics used in baghouses. Therefore, the determination of the stream dew point must consider the presence of SO_3. With information on the SO_3 content (ppm vol) and the moisture content (% vol) of the gas stream, the "acid" dew point temperature can be determined from Figure B.1-1. Figure B.1-1 provides dew points for two moisture levels, however, dew points can be estimated for other moisture values.

The SO_3 content of a combined gas stream is calculated by first converting the SO_3 concentration of

Table B.1-1 Dew Point Temperatures

Psychrometric Ratio	Gas Stream Temperatures (°F)										
	70	80	90	100	120	140	160	180	200	220	240
	Dew Point Temperatures (°F)										
0.000	0	0	0	0	0	0	0	0	0	0	0
0.005	54	58	61	65	70	76	81	86	89	93	96
0.010	62	65	68	71	77	82	86	90	94	97	100
0.015	68	72	75	77	82	86	90	94	97	100	103
0.020		77	80	82	87	91	94	97	100	103	106
0.025			85	87	91	94	98	101	103	106	109
0.030			89	91	95	98	100	104	107	109	111
0.035				95	98	101	104	107	109	110	114
0.040				98	101	104	107	109	111	114	116
0.045					104	107	109	112	114	116	118
0.050					107	109	112	114	116	118	120
0.055					109	112	114	116	118	120	122
0.060					111	114	116	118	120	122	124
0.065					114	116	118	120	122	124	125
0.070					116	118	120	122	123	125	130
0.075					118	120	122	124	125	130	150
0.080					119	122	123	125	130	140	170
0.085						123	125	130	143	168	182
0.090						124	130	140	162	180	205
0.095						128	140	165	180	205	225

Figure B.1-1. "Acid" dew points in stack gases.

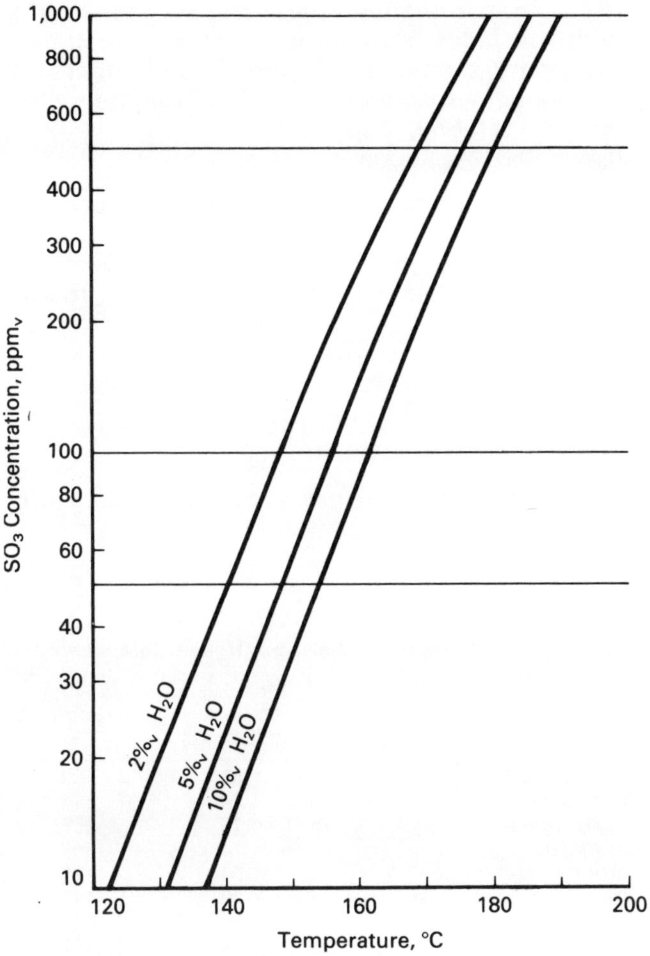

each individual stream to a lb-mole (lm) basis. The SO_3 content is calculated as follows:

$$S_{e1} \times \frac{1}{10^6} \times Q_{e1} \times \frac{\text{lb-mole}}{414 \text{ scf}} = S_{e1,lm}$$

where:
S_{e1} = SO_3 content of gas stream #1 (ppm vol)
$S_{e1,lm}$ = SO_3 content of gas stream #1 (lb-mole/min)

This is repeated for each separate gas stream. These are then added to obtain the total SO_3 content of the combined gas stream to the control device as follows:

$$S_{e1,lm} + S_{e2,lm} = \ldots = S_{e,lm}$$

$$S_{e,lm} \times \frac{414 \text{ scf}}{\text{lb-mole}} \times \frac{10^6}{Q_e} = S_e$$

where:
$S_{e,lm}$ = SO_3 content of combined gas stream (lb-mole/min)
S_e = SO_3 content of combined gas stream (ppm vol)

With information for the SO_3 content of the combined gas stream (S_e) and the moisture content of the combined gas stream (M_e), the acid dew point is determined from Figure B.1-1.

B.1.3 Particulate Matter Loading

Particulate matter concentrations usually are reported in grains per acf. The procedures below may be used to determine the particulate loading to a control device (in lbs/hr) when gas streams are combined.

$$W_{e1,g} \times Q_{e1,a} \times \frac{60 \text{ min}}{\text{hr}} \times \frac{\text{lb}}{7{,}000 \text{ gr}} = W_{e1,l}$$

where:
$W_{e1,g}$ = particulate loading for gas stream #1 (gr/acf)
$W_{e1,l}$ = particulate loading for gas stream #1 (lb/hr)

This is repeated for each gas stream and the results are added to obtain the particulate loading for the combined gas stream.

$$W_{e1,l} + W_{e2,l} + \ldots = W_{e,l}$$

where:
$W_{e,l}$ = particulate loading for combined gas stream (lb/hr)

The particulate loading of the combined gas stream can be converted to a concentration as follows:

$$W_{e,l} \times \frac{7{,}000 \text{ gr}}{\text{lb}} \times \frac{1 \text{ hr}}{60 \text{ min}} \times \frac{1}{Q_{e,a}} = W_{e,g}$$

where:
$W_{e,g}$ = particulate loading for combined gas stream (gr/acf)

B.1.4 Heat Content Calculation

The heat content of gas stream #1 (h_{e1}) can be determined from the heat of combustion of its components using the following equation:

$$h_{e1} = (0.01) \sum_{i=1}^{n} y_{e1,i} \times h_{e1,i}$$

where:
h_{e1} = heat content in gas stream #1 (Btu/scf)
$y_{e1,i}$ = volume percent of component "i" in gas stream #1 (% vol)
$h_{e1,i}$ = heat of combustion of component "i" in gas stream #1: see Table B.1-2 (Btu/scf)
n = number of components in gas stream #1

The heat content of a combined emission stream can be determined from the heat content of the individual emission streams as follows:

$$H_e = (0.01) \sum_{j=1}^{m} y_{ej} \times h_{ej}$$

where:

h_e = combined emission stream heat content (Btu/scf)

y_{ej} = volume percent of stream "j" in combined gas stream (% vol)

h_{ej} = heat content of stream "j" in combined gas stream: see previous discussion (Btu/scf)

m = number of individual gas streams in combined gas stream

Table B.1-2. Heats of Combustion and Lower Exposive Limit (LEL) Data for Selected Compounds[a]

Compound	LEL (ppmv)	Net Heat of Combustion[b,c] (Btu/scf)
Methane	50,000	892
Ethane	30,000	1,588
Propane	21,000	2,274
n-Butane	16,000	2,956
Isobutane	18,000	2,947
n-Pentane	15,000	3,640
Isopentane	14,000	3,631
Neopentane	14,000	3,616
n-Hexane	11,000	4,324
Ethylene	27,000	1,472
Propylene	20,000	2,114
n-Butene	16,000	2,825
1-Pentene	15,000	3,511
Benzene	13,000	3,527
Toluene	12,000	4,196
Xylene	11,000	1,877
Acetylene	25,000	1,397
Naphthalene	9,000	5,537
Methyl alcohol	60,000	751
Ethyl alcohol	33,000	1,419
Ammonia	160,000	356
Hydrogen sulfide	40,000	583

[a]Sources: *Steam/Its Generation and Use*. The Babcock & Wilcox Company. New York, NY. 1975.
Fire Hazard Properties of Flammable Liquids, Gases, Volatile Solids - 1977. National Fire Protection Association. Boston, MA. 1977.

[b]Lower heat of combustion.

[c]Based on 70°F and 1 atm.

Example Case

Calculate the heat content of an emission stream from a paper coating operation (gas stream #1) with the following composition data: methane (44 ppmv), toluene (73 ppmv), and others (4 ppmv). Let subscripts "1" and "2" denote methane and toluene, respectively.

$h_{e1} = (0.01)\,(y_{e1,1} \times h_{e1,1} + y_{e1,2} \times h_{e1,2})$

Convert the concentrations to volume percent basis:

Methane: $y_{e1,1} = 0.0048$ (assume "others" is equivalent to methane)

Toluene: $y_{e1,2} = 0.0073$

From Table B.1-2:

Methane: $h_{e1,1} = 892$ Btu/scf

Toluene: $h_{e1,2} = 4{,}196$ Btu/scf

Substituting these values in the above equation yields:

$h_{e1} = 0.35$ Btu/scf

Table B.1-3 Properties of Selected Organic Compounds[a]

Compound	Molecular Weight (lb/lb-mole)	Boiling Point (°F)
Acetone	58	133
Benzene	78	176
n-Butyl acetate	116	257
n-Butyl alcohol	74	243
Carbon tetrachloride	154	170
Chloroform	119	142
Cyclohexane	54	176
Ethyl acetate	88	171
Ethyl alcohol	46	173
Heptane	100	209
Hexane	86	156
Isobutyl alcohol	74	225
Isopropyl acetate	103	191
Isopropyl alcohol	60	181
Methyl acetate	74	135
Methyl alcohol	32	148
Methylene chloride	85	104
Methyl ethyl ketone	72	175
Methyl isobutyl ketone	100	244
Perchloroethylene	166	250
Toluene	92	231
Trichlorethylene	131	189
Trichlorotrifluoroethane	187	118
Xylene	106	281-292

[a]Source: *Chemical Engineer's Handbook*. Perry, R.H. and Chilton, C.H. (eds). Fifth Edition. McGraw-Hill Book Company. New York, NY. 1973.

Appendix B.2
Dilution Air Requirements Calculations

B.2.1 Dilution Air Calculations

The quantity of dilution air (Q_d) needed to decrease the heat content of the emission stream to h_d is given by the following equation:

$$Q_d = [(h_e/ h_d) - 1]Q_e \qquad (1)$$

where:

Q_d = dilution air flow rate, scfm

h_e = emission stream heat content before dilution, Btu/scf

h_d = emission stream heat content after dilution, Btu/scf

Q_e = emission stream flow rate before dilution, scfm

The concentrations of the various components and flow rate of the emission stream have to be adjusted after dilution as follows:

$$O_{2,d} = O_2\,(h_d/h_e) + 21\,[1 - (h_d/h_e)] \qquad (2)$$

$$M_{e,d} = M_e\,(h_d/h_e) + 2\,[1 - (h_d/h_e)] \qquad (3)$$

$$Q_{e,d} = Q_e\,(h_e/ h_d) \qquad (4)$$

where:

$O_{2,d}$ = oxygen content of diluted emission stream, vol %

$M_{e,d}$ = moisture content of diluted emission stream, vol %

$Q_{e,d}$ = flow rate of the diluted emission stream, scfm

The factor 21 in Equation 2 denotes the volumetric percentage of oxygen in air and the factor 2 in Equation 3 is the volumetric percentage of moisture in air at 70°F and 80 percent humidity.

After dilution, the HAP emission stream characteristics are redesignated as follows:

$O_2 = O_{2,d}$ = __________ %
$M_e = M_{e,d}$ = __________ %
$h_e = h_d$ = __________ Btu/scf
$Q_e = Q_{e,d}$ = __________ scfm

Appendix C.2 is a worksheet for calculating dilution air requirements.

Appendix C.1
HAP Emission Stream Data Form

HAP EMISSION STREAM DATA FORM*

Company ______________________ Plant Contact ______________________

Location (Street) ______________________ Telephone No. ______________________

(City) ______________________ Agency Contact ______________________

(State, Zip) ______________________ No. of Emission Streams Under Review ______________________

A. Emission Stream Number/Plant Identification ______________________

B. HAP Emission Source	(a) ______	(b) ______	(c) ______
C. Source Classification	(a) ______	(b) ______	(c) ______
D. Emission Stream HAP's	(a) ______	(b) ______	(c) ______
E. HAP Class and Form	(a) ______	(b) ______	(c) ______
F. HAP Content (1,2,3)**	(a) ______	(b) ______	(c) ______
G. HAP Vapor Pressure (1,2)	(a) ______	(b) ______	(c) ______
H. HAP Solubility (1,2)	(a) ______	(b) ______	(c) ______
I. HAP Adsorptive Prop. (1,2)	(a) ______	(b) ______	(c) ______
J. HAP Molecular Weight (1,2)	(a) ______	(b) ______	(c) ______

K. Moisture Content (1,2,3) ______	P. Organic Content (1)*** ______
L. Temperature (1,2,3) ______	Q. Heat/O_2 Content (1) ______
M. Flow Rate (1,2,3) ______	R. Particulate Content (3) ______
N. Pressure (1,2) ______	S. Particle Mean Diam. (3) ______
O. Halogen/Metals (1,2) ______	T. Drift Velocity/SO_3 (3) ______

U. Applicable Regulation(s) ______________________

V. Required Control Level ______________________

W. Selected Control Methods ______________________

*The data presented are for an emission stream (single or combined streams) prior to entry into the selected control method(s). Use extra forms if additional space is necessary (e.g., more than three HAP's). and note this need.

**The numbers in parentheses denote what data should be supplied depending on the data in steps "C" and "E":

1 = organic vapor process emission
2 = inorganic vapor process emission
3 = particulate process emission

***Organic emission stream combustibles less HAP combustibles shown on Lines D and F.

Appendix C.2
Calculation Sheet for Dilution Air Requirements

Dilution air flow rate:

$Q_d = [(h_e/ h_d) - 1]Q_e$

Q_d = ________ scfm

Diluted emission stream characteristics:

$O_{2,d} = O_2 (h_d/h_e) + 21 [1 - (h_d/h_e)]$

$O_{2,d}$ = ________ %

$M_{e,d} = M_e (h_d/h_e) + 2 [1 - (h_d/h_e)]$

$M_{e,d}$ = ________ %

$Q_{e,d} = Q_e (h_e/ h_d)$

$Q_{e,d}$ = ________ scfm

Redesignate emission stream characteristics:

$O_2 = O_{2,d}$ = ________ %

$M_e = M_{e,d}$ = ________ %

$h_e = h_d$ = ________ Btu/scf

$Q_e = Q_{e,d}$ = ________ scfm

Appendix C.3
Calculation Sheet for Thermal Incineration

4.2.1 Data Required

HAP emission stream characteristics:[a]

1. Maximum flow rate, Q_e = ________ scfm
2. Temperature, T_e = ________ °F
3. Heat content, h_e = ________ Btu/scf
4. Oxygen content,[b] O_2 = ________ %
5. Moisture content, M_e = ________ %
6. Halogenated organics: Yes ________ No ________

Required destruction efficiency, DE = ________ %

In the case of a permit review, the following data should be supplied by the applicant:

Thermal incinerator system variables at standard conditions (70°F, 1 atm):

1. Reported destruction efficiency, $DE_{reported}$ = ________ %
2. Temperature of emission stream entering the incinerator,

   T_e = ________ °F (if no heat recovery);

   T_{he} = ________ °F (if a heat exchanger is employed)
3. Combustion temperature, T_c = ________ °F
4. Residence time, t_r = ________ sec
5. Maximum emission stream flow rate, Q_e = ________ scfm
6. Excess air, ex = ________ %

---

[a] If dilution air is added to the emission stream upon exit from the process, the data required are the resulting characteristics after dilution.

[b] The oxygen content depends on the oxygen content of the organic compounds (fixed oxygen) and the free oxygen in the emission stream. Since emission streams treated by thermal incineration are generally dilute VOC and air mixtures, the fixed oxygen in the organic compounds can be neglected.

7. Fuel heating value , h_f = ________ Btu/scf (assume natural gas)

8. Supplementary heat requirement, H_f = ________ Btu/min

9. Combustion chamber volume, V_c = ________ ft^3

10. Flue gas flow rate, Q_{fg} = ________ scfm

11. Heat exchanger surface area (if a heat exchanger is employed),

    A = ________ ft^2

4.2.2 Pretreatment of the Emissions Stream: Dilution Air Requirements

Typically, dilution will not be required. However, if the emission stream heat content (h_e) is greater than 13 Btu/scf with oxygen concentration greater than 16 percent, see Appendix C.2 where a blank calculation sheet for determining dilution air requirements is provided.

4.2.3 Thermal Incinerator System Design Variables

Based on the required destruction efficiency (DE), select appropriate values for T_c and t_r from Table 4-1.

T_c = ________ °F

t_r = ________ sec

For a permit evaluation, if the applicant's values for T_c and t_r are sufficient to achieve the required DE (compare the reported values with the values presented in Table 4-1), proceed with the calculations. If the applicant's values for T_c and t_r are not sufficient, the applicant's design is unacceptable. The reviewer may then use the values for T_c and t_r from Table 4-1.

T_c = ________ °F

t_r = ________ sec

(Note: If DE is less than 98 percent, obtain information from literature and incinerator vendors to determine appropriate values for T_c and t_r.)

4.2.4 Determination of Incinerator Operating Variables

4.2.4.1 Supplementary Heat Requirements

1. For dilute emission streams that require no additional combustion air:

    a. Use Figure 4-2:

$$H_f = (H_f/Q_e)_{figure}\, Q_e$$

H_f = ________ Btu/min

or

b. Use Equation 4.2-1:

$$H_f = 1.1\ h_f \left[\frac{Q_e(1 + 0.002\ M_e)\ [Cp_{air}\ (T_c\text{-}T_r) - Cp_{air}\ (T_{he}\text{-}T_r) - h_e]}{h_f - 1.4\ Cp_{air}\ (T_c - T_r)}\right]$$

The values for the parameters in this equation an be determined as follows:

Q_e, h_e, M_e — Input data.

h_f — Assume a value of 882 Btu/scf if no other information is available.

Cp_{air} — Assume a value of 0.0190 Btu/scf-°F if no other information is available.

T_c — Obtain value from Table 4-1 or from permit applicant.

T_{he} — Use the following equation if the value for T_{he} is not specified:

$$T_{he} = (HR/100)\ T_c + [1 - (HR/100)]\ T_e$$

where HR = heat recovery in the heat exchanger (percent). Assume a value of 50 percent for HR if no other information is available.

T_r — 70°F

H_f = ________ Btu/min

If H_f is less than 5 Btu/min, redefine H_f = 5 Btu/min.

2. For emission streams that are not dilute and require additional combustion air:

Use Figure 4-3 to obtain a conservative estimate:

$$H_f = (H_f/Q_e)_{figure}\ Q_e$$

H_f = ________ Btu/min

4.2.4.2 Flue Gas Flow Rate

1. For dilute emission streams, use Equation 4.2-2:

   $Q_{fg} = Q_e + Q_f + Q_c$

   where:

   $Q_c = 0$

   and

   $Q_f = H_f/h_f$

   $Q_{fg} =$ ________ scfm

2. For emission streams that require additional combustion air, use the following equation to calculate Q_c:

   $Q_c = [(0.01\ H_e + 9.4\ Q_f)\ (1 + 0.01\ ex) - 0.0476\ O_2\ Q_e]$

   $H_e = Q_e h_e$

   Assume ex = 18 percent if no other information is available.

   $Q_c =$ ________ scfm

   Then use Equation 4.2-2 to calculate Q_{fg}:

   $Q_{fg} =$ ________ scfm

4.2.5 Combustion Chamber Volume

a. Use Equation 4.2-4 to Convert Q_{fg} (standard conditions) to $Q_{fg,a}$ (actual conditions):

   $Q_{fg,a} = Q_{fg}\ [(T_c + 460)/530]$

   (Note: Pressure effects are negligible.)

   $Q_{fg,a} =$ ________ acfm

b. Use Equation 4.2-5 to calculate combustion chamber volume:

   $V_c = [(Q_{fg,a}/60)\ t_r]\ 1.05$

   Obtain value for t_r from Table 4-1 or from permit applicant.

   $V_c =$ ________ ft^3

If V_c is less than 36 ft^3 (minimum commercially available size),

$V_c = 36\ ft^3$

4.2.6 Heat Exchanger Size

1. For dilute emission streams that do not require additional combustion air:

   a. Use Figure 4-4:

   $A =$ ________ ft^2

   b. Use Equation 4.2-6:

   $A = [60\ Q_e\ (1 + 0.002\ M_e)\ Cp_{air}\ (T_{he} - T_e)]/U\ \Delta T_{LM}$

   The values for the parameters in this equation can be determined as follows:

Q_e, Cp_{air}, T_{he}, M_e	As specified for Equation 4.2-1.
T_e	Input data.
U	Use a value of 4 Btu/hr-ft^2-°F unless the inquirer/ applicant has provided a value.
T_c	As specified for Equation 4.2-1.
ΔT_{LM}	Calculate ΔT_{LM} using the following expression: $\Delta T_{LM} = T_c - T_{he}$; $\Delta T_{LM} =$ ________ °F

   Heat exchanger surface area:

   $A =$ ________ ft^2

2. For emission streams that are not dilute and require additional combustion air:

   Use Figure 4-5 :

   $A = (A/Q_e)_{figure}\ Q_e$

   $A =$ ________ ft^2

4.2.7 Evaluation of Permit Application

Compare the calculated values and reported values using Table 4-2. The combustion volume (V_c) is calculated from flue gas flow rate (Q_{fg}) and Q_{fg} is determined by emission stream flow rate (Q_e), supplementary fuel flow rate (Q_f), and combustion air requirement (Q_c). Therefore, if there are differences between the calculated and reported values for V_c and Q_{fg}, these are dependent on the differences between the calculated and reported values for Q_c and Q_f.

If the calculated and reported values are different, the differences may be due to the assumptions involved in the calculations. Discuss the details of the design and operation of the system with the applicant.

If the calculated and reported values are not different, then the design and operation of the system can be considered appropriate based on the assumptions employed in this handbook.

Table 4-2. Comparison of Calculated Values and Values Supplied by the Permit Applicant for Thermal Incineration

	Calculated Value	Reported Value
Supplementary heat requirement, H_f	. . .	. . .
Supplementary fuel flow rate, Q_f	. . .	. . .
Flue gas flow rate, Q_{fg}	. . .	. . .
Combustion chamber size, V_c	. . .	. . .
Heat exchanger surface area, A	. . .	. . .

Appendix C.4
Calculation Sheet for Catalytic Incineration

4.3.1 Data Required

HAP emission stream characteristics:[a]

1. Maximum flow rate, Q_e = ________ scfm
2. Temperature, T_e = ________ °F
3. Heat content, h_e = ________ Btu/scf
4. Oxygen content[b], O_2 = ________ %
5. Moisture content, M_e = ________ %

Required destruction efficiency, DE ________ %

In the case of a permit review, the following data should be supplied by the applicant:

Catalytic incinerator system variables at standard conditions (70°F, 1 atm):

1. Reported destruction efficiency, $DE_{reported}$ = ________ %
2. Temperature of emission stream entering the incinerator,

   T_e = ________ °F (if no heat recovery),

   T_{he} = ________ °F (if emission stream is preheated)
3. Temperature of flue gas leaving the catalyst bed,

   T_{co} = ________ °F

---

[a] If dilution air is added to the emission stream upon exit from the process, the data required are the resulting characteristics after dilution.

[b] The oxygen content depends on the oxygen content of the organic compounds (fixed oxygen) and the free oxygen in the emission stream. Since emission streams treated by catalytic incineration are generally dilute VOC and air mixtures, the fixed oxygen in the organic compounds can be neglected.

4. Temperature of combined gas stream (emission stream + supplementary fuel combustion products) entering the catalyst bed,[a]

   T_{ci} = ________ °F

5. Space velocity, SV = ________ hr^{-1}

6. Supplementary heat requirement, H_f = ________ Btu/min

7. Flow rate of combined gas stream entering the catalyst bed,

   Q_{com} = ________ scfm

8. Combustion air flow rate, Q_c = ________ scfm

9. Excess air, ex = ________ %

10. Catalyst bed requirement, V_{bed} = ________ ft^3

11. Fuel heating value, h_f = ________ Btu/scf

12. Heat exchanger surface area (if a heat exchanger is employed),

    A = ________ ft^2

4.3.2 Pretreatment of the Emission Stream: Dilution Air Requirements

For emission streams treated by catalytic incineration, dilution air typically will not be required. However, if the emission stream heat content is greater than 10 Btu/scf for air + VOC mixtures or if the emission stream heat content is greater than 15 Btu/scf for inert + VOC mixures, dilution air is necessary. For emission streams that cannot be characterized as air + VOC or inert + VOC mixtures, assume that dilution air will be required if the heat content is greater than 12 Btu/scf. In such cases, refer to Appendix C.2 where a blank calculation sheet for determining dilution air requirements is provided.

4.3.3 Catalytic Incinerator System Design Variables

Based on the required destruction efficiency (DE), specify the appropriate ranges for T_{ci} and T_{co} and select the value for SV from Table 4-3.

T_{ci} (minimum) = 600°F

T_{co} (minimum) = 1,000°F

T_{co} (maximum) = 1,200°F

SV = ________ hr^{-1}

---

[a]If no supplementary fuel is used, the value for this variable will be the same as that for the emission stream.

In a permit review, determine if the reported values for T_{ci}, T_{co}, and SV are appropriate to achieve the required destruction efficiency. Compare the applicant's values with the values in Table 4-3 and check if:

$$T_{ci} \text{ (applicant)} \geq 600°F \text{ and } 1{,}200°F \geq T_{co} \text{ (applicant)} \geq 1{,}000°F$$

and

$$SV \text{ (applicant)} \leq SV \text{ (Table 4-3)}$$

If the reported values are appropriate, proceed with the calculations. Otherwise, the applicant's design is considered unacceptable. The reviewer may then wish to use the values in Table 4-3.

4.3.4 Determination of Incinerator Operating Parameters

4.3.4.1 Supplementary Heat Requirements

1. For dilute emission streams that require no additional combustion air:

   a. Use Equation 4.3-1 to determine if $T_{ci} = 600°F$ from Table 4-3 is sufficient to ensure an adequate overall reaction rate without damaging the catalyst, i.e., check if T_{co} falls in the interval 1,000° - 1,200°F:

   $$T_{co} = 600 + 50\ h_e$$

   T_{co} = ________ °F

   If T_{co} falls in the interval 1,000° - 1,200°F, proceed with the calculations. If T_{co} is less than 1,000°F, assume T_{co} is equal to 1,000°F and use Equation 4.3-2 to determine an appropriate value for T_{ci}; and then proceed with the calculations:

   $$T_{ci} = 1{,}000 - 50\ h_e$$

   T_{ci} = ________ °F

   (Note: If T_{co} is greater than 1,200°F, decline in catalyst activity may occur due to exposure to high temperatures.)

   b. Use Figure 4-7 to determine supplementary heat requirements:

   $$H_f = (H_f/Q_e)_{figure}\ Q_e$$

   H_f = ________ Btu/min

   c. Use Equation 4.3-3 to determine supplementary heat requirements:

$$Hf = 1.1\, h_f\, Q_e\, (1 + 0.002\, M_e) \left[\frac{[Cp_{air}\,(T_{ci}\text{-}T_r) - Cp_{air}\,(T_{he}\text{-}T_r)]}{h_f - 1.4\, Cp_{air}\,(T_{ci} - T_r)}\right]$$

The values for the variables in this equation can be determined as follows:

Q_e, M_e, T_e — Input data.

h_f — Assume a value of 882 Btu/scf (for natural gas) if no other information is available.

C_{pair} — Assume a value of 0.0190 Btu/scf-°F if no other information is available.

T_{ci} — Obtain value from part a above or from permit applicant.

T_{he} — For no heat recovery case, $T_{he} = T_e$. For heat recovery case, use the following equation if the value for T_{he} is not specified:

$$T_{he} = (HR/100)T_{co} + [1 - (HR/100)]\, T_e$$

where HR = heat recovery in the heat exchanger (percent). Assume a value of 50 percent for HR if no other information is available.

T_r — 70°F

H_f = __________ Btu/min

2. For emission streams that are not dilute and require additional combustion air:

Use Figure 4-8 to obtain a conservative estimate:

$$H_f = (H_f/Q_e)\text{figure}\; Q_e$$

H_f = __________ Btu/min

4.3.4.2. Flow Rate of Combined Gas Stream Entering the Catalyst Bed

1. For dilute emission streams that require no additional combustion air, use Equations 4.3-4 and -5:

$$Q_{com} = Q_e + Q_f + Q_c$$

$$Q_f = H_f/h_f$$

Q_f = __________ scfm

Q_{com} = __________ scfm

2. For emission streams that require additional combustion air, use the following equation to calculate Q_c:

$$Q_c = [(0.01\ h_e\ Q_e + 9.4\ Q_f)\ (1 + 0.01\ ex) - 0.0476\ O_2\ Q_e]$$

Q_c = __________ scfm

Then use Equation 4.3-4 to calculate Q_{com}:

Q_{com} = __________ scfm

4.3.4.3 Flow Rate of Flue Gas Leaving the Catalyst Bed

a. Use the result from the previous calculation:

$$Q_{fg} = Q_{com}$$

Q_{fg} = __________ scfm

If Q_{fg} is less than 500 scfm, define Q_{fg} as 500 scfm.

b. Use Equation 4.3-6 to calculate $Q_{fg,a}$:

$$Q_{fg,a} = Q_{fg}\ [(T_{co} + 460)/530]$$

$Q_{fg,a}$ = __________ acfm

4.3.5 Catalyst Bed Requirement

Use Equation 4.3-7:

$$V_{bed} = 60\ Q_{com}/SV$$

V_{bed} = __________ ft^3

4.3.6 Heat Exchanger Size (for Systems with Recuperative Heat Exchange Only)

1. For dilute emission streams that do not require additional combustion air:

a. Use Figure 4-9 (line 1):

$$A = (A/Q_e)_{figure}\ Q_e$$

A = __________ ft^2

b. Use Equation 4.3-8:

$$A = [60\ Q_e\ (1 + 0.002\ M_e)\ Cp_{air}\ (T_{he}\text{-}T_e)/U\Delta T_{LM}]$$

The values for the parameters in this equation can be determined as follows:

Q_e, Cp_{air}, T_{he}, M_e, h_e	As specified for Equations 4.3-1 and -3.
T_e	Input data.
U	Use a value of 4 Btu/hr-ft^2-°F unless the inquirer/applicant has provided a value.
T_{co}	As calculated in Step 1 of 4.3.4.1: T_{co} = ________ °F
ΔT_{LM}	Calculate ΔT_{LM} using the following expression: $\Delta T_{LM} = T_{co} - T_{he}$; ΔT_{LM} = ________ °F

Heat exchanger surface area:

A = ________ ft^2

2. For emission streams that are not dilute and require additional combustion air:

Use Figure 4-9 (line 2):

$$A = (A/Q_e)_{figure}\ Q_e$$

A = ________ ft^2

4.3.7 Evaluation of Permit Application

Compare the calculated values and the values supplied by the applicant using Table 4-4.

If the calculated values for H_f, Q_c, Q_{com}, V_{bed}, and A differ from the applicant's values, the differences may be due to the assumptions involved in the calculations. Discuss the details of the design and operation of the system with the applicant.

If the calculated and reported values are not different, then the design and operation of the system can be considered appropriate based on the assumptions employed in this handbook.

Table 4-4. Comparison of Calculated Values and Values Supplied by the Permit Applicant for Catalytic Incineration

	Calculated Value	Reported Value
Supplementary heat requirement, H_f	. . .	. . .
Supplementary fuel flow rate, Q_f	. . .	. . .
Combustion air flow rate, Q_{fc}	. . .	. . .
Flue gas stream flow rate, C_{com}	. . .	. . .
Catalyst bed volume, V_{bed}	. . .	. . .
Heat exchanger surface area, (if recuperative heat recovery is used), A	. . .	. . .

Appendix C.5
Calculation Sheet for Flares

4.4.1 Data Required

HAP emission stream characteristics:

1. Expected emission stream flowrate, Q_e = ________ scfm

2. Emission stream temperature, T_e = ________ °F

3. Heat content, h_e = ________ Btu/scf

4. Mean molecular weight of emission stream MW_e = ________ lb/lb-mole

Flare tip diameter, D_{tip} = ________ in

Required destruction efficiency, DE = ________ %

In the case of a permit review, the following data should be supplied by the applicant:

Flare system design parameters at standard conditions (70°F, 1 atm):

1. Flare tip diameter, D_{tip} = ________ in

2. Expected emission stream flowrate, Q_e = ________ scfm

3. Emission stream heat content, h_e = ________ Btu/scf

4. Temperature of emission stream, T_e = ________ °F

5. Mean molecular weight of emission stream, MW_e = ________ lb/lb-mole

6. Steam flowrate, Q_s = ________ lb/min

7. Flare gas exit velocity, U_{flg} = ________ ft/sec

8. Supplementary fuel flow rate,[a] Q_f = ________ scfm

9. Supplementary fuel heat content,[a] h_f = ________ Btu/scf

---

[a] This information is needed if the emission stream heat content is less than 300 Btu/scf.

10. Temperature of flare gas,[b] T_{flg} = __________ °F

11. Flare gas flowrate,[b] Q_{flg} = __________ scfm

12. Flare gas heat content,[b] h_{flg} = __________ Btu/scf

4.4.2 Determination of Flare Operating Variables

Based on studies conducted by EPA, relief gases having heating values less than 300 Btu/scf are not assured of achieving 98 percent destruction efficiency when they are flared in steam- or air-assisted flares.[c]

In a permit review case, if h_e is below 300 Btu/scf and no supplementary fuel is used, then the application is rejected. The reviewer may then wish to proceed with the calculations below. If h_e is equal to or above 300 Btu/scf, then the reviewer should skip to Section 4.4.2.3.

4.4.2.1 Supplementary Fuel Requirements

For emission streams with heat contents less than 300 Btu/scf, additional fuel is required. Use Equation 4.4-1 to calculate natural gas requirements:

$$Q_f = [(300 - h_e)\ Q_e]/582$$

Q_f = __________ scfm

4.4.2.2 Flare Gas Flow Rate and Heat Content

a. Use Equation 4.4-2 to calculate the flare gas flow rate:

$$Q_{flg} = Q_e + Q_f$$

Q_{flg} = __________ scfm

b. Determine the flare gas heat content as follows:

$$h_{flg} = 300 \text{ Btu/scf if } Q_f > 0$$

$$h_{flg} = h_e \text{ if } Q_f = 0$$

h_{flg} = __________ Btu/scf

4.4.2.3 Flare Gas Exit Velocity

a. Use Table 4-5 to calculate U_{max}:

---

[b] If no auxiliary fuel is added, the value for this variable will be the same as that for the emission stream.

[c] For unassisted flares, the lower limit is 200 Btu/scf.

If $300 \leq h_{flg} < 1{,}000$, use the following equation:

$$U_{max} = 3.28\,[10^{(0.00118 h_{flg} + 0.908)}]$$

U_{max} = __________ ft/sec

If $h_{flg} \geq 1{,}000$ Btu/scf, $U_{max} = 400$ ft/sec

b. Use Equation 4.4-3 to calculate U_{flg}:

$$U_{flg} = 3.06\, Q_{flg,a}/(D_{tip})^2$$

where $Q_{flg,a}$ is given by Equation 4.4-4:

$$Q_{flg,a} = [Q_{flg}\,(T_{flg} + 460)]/530$$

See Appendix B.7, reference 8, for calculating T_{flg}.

$Q_{flg,a}$ = __________ acfm

U_{flg} = __________ ft/sec

c. Compare U_{flg} and U_{max}:

If $U_{flg} \leq U_{max}$, the desired destruction efficiency level of 98 percent can be achieved. (Note: U_{flg} should exceed 0.03 ft/sec for flame stability.) If $U_{flg} > U_{max}$, 98 percent destruction efficiency cannot be achieved. When evaluating a permit, reject the application in such a case.

4.4.2.4 Steam Requirements

a. Assume that the amount of steam required is 0.4 lb steam/lb flare gas. Use Equation 4.4-5 to calculate Q_s:

$$Q_s = 1.03 \times 10^{-3} \times Q_{flg} \times MW_{flg}$$

Q_s = __________ lb/min

4.4.3 Evaluation of Permit Application

Compare the calculated and reported values using Table 4-6. If the calculated values of Q_f, U_{flg}, Q_{flg}, and Q_s are different from the reported values for these variables, the differences may be due to the assumptions (e.g., heating value of fuel, ratio of steam to flare gas, etc.) involved in the calculations. Discuss the details of the design and operation of the system with the applicant. If the calculated and reported values are not different, then the operation of the system can be considered appropriate based on the assumptions employed in the handbook.

Table 4-6. Comparison of Calculated Values and Values Supplied by the Permit Applicant for Flares

	Calculated Value	Reported Value
Supplementary fuel flow rate, Q_f	. . .	. . .
Flare gas exit velocity, U_{flg}	. . .	. . .
Flare gas flow rate, Q_{flg}	. . .	. . .
Steam flow rate, Q_s	. . .	. . .

Appendix C.6
Calculation Sheet for Carbon Adsorption

4.6.1 Data Required

HAP Emission stream characteristics:

1. Maximum flow rate, Q_e = ________ scfm
2. Temperature, T_e = ________ °F
3. Relative humidity, R_{hum} = ________ %
4. HAP = ________
5. Maximum HAP content, HAP_e = ________ ppmv

Required removal efficiency, RE = ________ %

In the case of a permit review, the following data should be supplied by the applicant:

Carbon adsorber (fixed-bed) system variables at standard conditions (70°F, 1 atm):

1. Reported removal efficiency, $RE_{reported}$ = ________ %
2. HAP content, HAP_e = ________ ppmv
3. Emission stream flow rate, Q_e = ________ scfm
4. Adsorption capacity of carbon bed,

   AC = ________ lb HAP/100 lb carbon
5. Number of beds = ________
6. Amount of carbon required, C_{req} = ________ lb
7. Cycle time for adsorption, θ_{ad} = ________ hr
8. Cycle time for regeneration, θ_{reg} = ________ hr
9. Emission stream velocity through the bed, U_e = ________ ft/min
10. Bed depth, Z_{bed} = ________ ft
11. Bed diameter, D_{bed} = ________ ft
12. Steam ratio, St = ________ lb steam/lb carbon

4.6.2 Pretreatment of the Emission Stream

4.6.2.1 Cooling

$$T_e = \underline{\hspace{2cm}} \ ^\circ F$$

If the temperature of the emission stream is significantly higher than 100°F, a heat exchanger is needed to cool it to 100°F. Refer to Appendix B.5, reference 8, for the calculation procedure.

4.6.2.2 Dehumidification

$$R_{hum} = \underline{\hspace{2cm}} \ \%$$

If the relative humidity level is above 50 percent, a condenser is required to cool and condense the water vapor in the emission stream. Refer to Section 4.8 for more details.

4.6.2.3 High VOC Concentrations

$$HAP_e = \underline{\hspace{2cm}} \ \text{ppmv}$$

If flammable vapors are present in the emission stream, VOC content will be limited to below 25 percent of the LEL.

$$LEL = \underline{\hspace{2cm}} \ \text{ppmv (from Table B.1-1)}$$

$$25\% \text{ of } LEL = 0.25 \times LEL \text{ (ppmv)} = \underline{\hspace{2cm}} \ \text{ppmv}$$

The maximum practical inlet concentration for carbon beds is about 10,000 ppmv. If HAP_e is greater than 10,000 ppmv, carbon adsorption may not be applicable.

4.6.3 Carbon Adsorption System Design Variables

a. Use Equation 4.6-1 to calculate the required outlet HAP concentration:

$$HAP_o = HAP_e \ (1 - 0.01 \ RE)$$

$$HAP_o = \underline{\hspace{2cm}} \ \text{ppmv}$$

b. Specify the appropriate values of θ_{ad}, θ_{reg}, and St from Table 4-7.

$$\theta_{ad} = \underline{\hspace{2cm}} \ \text{hr}$$

$$\theta_{reg} = \underline{\hspace{2cm}} \ \text{hr}$$

$$St = \underline{\hspace{2cm}} \ \text{lb steam/lb carbon}$$

4.6.4 Determination of Carbon Adsorber System Variables

4.6.4.1 Carbon Requirements

a. Use Equation 4.6-2:

$$C_{req} = 2 \times 1.55\ 10^{-5}\ N\theta_{ad}\ Q_e\ (HAP_e - HAP_o)\ MW_{HAP}/AC$$

Assume N = 2

Obtain MW_{HAP} from Table B.1-2 or reference 18. If no data are available, use a conservative value of 5 lb HAP/100 lb carbon.

C_{req} = __________ lb

b. Use Figure 4-13 to obtain (C_{req}/Q_e):

$$C_{req} = (C_{req}/Q_e)_{figure}\ Q_e$$

C_{req} = __________ lb

4.6.4.2 Carbon Adsorber Size

a. Use Equation 4.6-3 to calculate A_{bed}:

$$A_{bed} = Q_{e,a}/U_e$$

Calculate $Q_{e,a}$ using Equation 4.6-4:

$$Q_{e,a} = Q_e\ [(T_e + 460)/530]$$

$Q_{e,a}$ = __________ acfm

Assume U_e = 100 ft/sec

A_{bed} = __________ ft^2

b. Use Equation 4.6-5 to calculate D_{bed}:

$$D_{bed} = 1.13\ (A_{bed})^{0.5}$$

D_{bed} = __________ ft

c. Use Equation 4.6-6 to calculate volume of carbon per bed:

$$V_{carbon} = (C_{req}/N)/\rho_{bed}$$

Assume ρ_{bed} = 30 lb/ft^3

V_{carbon} = __________ ft^3

d. Use Equation 4.6-7 to calculate Z_{bed}:

$$Z_{bed} = V_{carbon}/A_{bed}$$

Z_{bed} = __________ ft

Note: If Q_e is greater than about 20,000 scfm, three or more carbon beds may need to be used.

4.6.4.3 Steam Required for Regeneration

a. Use Equation 4.6-8 to calculate steam requirements:

$$Q_s = [St \times C_{req}/(\theta_{reg} - \theta_{dry\text{-}cool})]/60$$

Assume $\theta_{dry\text{-}cool} = 0.25$ hrs.

Q_s = __________ lb/min

b. Use Figure 4-14:

Q_s = __________ lb/min

Calculate Q_s/A_{bed}:

Q_s/A_{bed} = __________ lb steam/min-ft^2

If Q_s/A_{bed} is greater than 4 lb steam/min-ft^2, fluidization of the carbon bed may occur.

4.6.4.4 Condenser

a. Use Equation 4.6-10 to calculate H_{load}:

$$H_{load} = 1.1 \times 60 \times Q_s\ [\lambda + \underline{C}p_w(T_{sti} - T_{sto})]$$

Obtain λ and $\underline{C}p_w$ from reference 19 based on the values assumed for T_{sti} and T_{sto}.

H_{load} = __________ Btu/hr

b. Use Equation 4.6-9 to calculate A_{con}:

$$A_{con} = H_{load}/U\Delta T_{LM}$$

Assume U = 150 Btu/hr-ft^2-°F if no other data are available.

$$\Delta T_{LM} = \left[\frac{(T_{sti} - T_{wo}) - (T_{sto} - T_{wi})}{\ln[(T_{sti} - T_{wo}) - (T_{sto} - T_{wi})]}\right]$$

where $T_{wi} = 80°F$ and $T_{wo} = 130°F$.

ΔT_{LM} = __________ °F

A_{con} = __________ ft²

c. Use Equations 4.6-11 and -12 to calculate Q_w:

$$Q_{cool,w} = H_{load}/[\underline{C}p_w(T_{wo} - T_{wi})]$$

$Q_{cool,w}$ = __________ lb/hr

$$Q_w = 0.002\ Q_{cool,w}$$

Q_w = __________ gal/min

4.6.4.5 Recovered Product

a. Use Equation 4.6-13 to calculate Q_{rec}:

$$Q_{rec} = 1.55 \times 10^{-9} \times Q_e \times HAP_e \times RE \times MW_{HAP}$$

Q_{rec} = __________ lb/hr

4.6.5 Evaluation of Permit Application

Compare the results from the calculations and the reported values using Table 4-8.

If the calculated values of C_{req}, D_{bed}, Z_{bed}, Q_s, A_{con}, Q_w, and Q_{roc}, are different from the reported values, the differences may be due to the assumptions involved in the calculations. Discuss the details of the design and operation of the system with the applicant.

If the calculated values agree with the reported values, then the design and operation of the proposed carbon adsorber system may be considered appropriate based on the assumptions made in this handbook.

Table 4-8. Comparison of Calculated Values and Values Supplied by the Permit Applicant for Carbon Adsorption

	Calculated Value	Reported Value
Carbon requirement, C_{req}	...	...
Bed diameter, D_{bed}	...	...
Bed depth, Z_{bed}	...	...
Steam rate, Q_s	...	...
Condenser surface area, A_{con}	...	...
Cooling water rate, Q_w	...	...
Recovered product, Q_{rec}	...	...

Appendix C.7
Calculation Sheet for Absorption

4.7.1 Data Required

HAP emission stream characteristics:

1. Maximum flow rate, Q_e = ________ scfm

2. Temperature, T_e = ________ °F

3. HAP = ________

4. HAP concentration, HAP_e = ________ ppmv

5. Pressure, P_e = ________ mm Hg

Required removal efficiency, RE = ________ %

In the case of a permit review, the following data should be supplied by the applicant:

Absorption system variables at standard conditions (70°F, 1 atm):

1. Reported removal efficiency, $RE_{reported}$ = ________ %

2. Emission stream flow rate, Q_e = ________ scfm

3. Temperature of emission stream, T_e = ________ °F

4. HAP = ________

5. HAP concentration, HAP_e = ________ ppmv

6. Solvent used = ________

7. Slope of the equilibrium curve, m = ________

8. Solvent flow rate, L_{gal} = ________ gal/min

9. Density of the emission stream, ρ_G = ________ lb/ft^3

10. Schmidt No. for the HAP/emission stream and HAP/solvent systems:

    Sc_G = ________
    Sc_L = ________

    (Refer to Appendix B.9, reference 8, or reference 12 for definition and calculation of Sc_G and Sc_L)

11. Properties of the solvent:

Density, ρ_L = ________ lb/ft^3

Viscosity, μ_L = ________ centipoise

12. Type of packing used = ________

13. Packing constants:

a = ________ b = ________ c = ________ d = ________

ϵ = ________ Y = ________ s = ________ g = ________

r = ________

14. Column diameter, D_{column} = ________ ft

15. Tower height, (packed) Ht_{column} = ________ ft

16. Pressure drop, ΔP_{total} = ________ in H_2O

4.7.3 Determination of Absorber System Design and Operating Variables

4.7.3.1 Solvent Flow Rate

a. Assume a value of 1.6 for AF.

Determine "m" from the equilibrium data for the HAP/solvent system under consideration (see references 18, 22, and 23 for equilibrium data).

m = ________

Use Equation 4.7-3:

Q_e = ________ scfm

$G_{mol} = 0.155\ Q_e$

G_{mol} = ________ lb-moles/hr

b. Use Equation 4.7-2:

$L_{mol} = 1.6\ m\ G_{mol}$

L_{mol} = ________ lb-moles/hr

c. Use Equation 4.7-4:

$L_{gal} = 0.036\ L_{mol}$

$L_{gal} =$ ________ gal/min

4.7.3.2 Column Diameter

a. Use Figure 4-17:

Calculate the abscissa (ABS):

$MW_{solvent} =$ ________ lb/lb-mole

$L = L_{mol} \times MW_{solvent}$

$L =$ ________ lb/hr

$MW_e =$ ________ lb/lb-mole

$G = G_{mol} \times MW_e$

$G =$ ________ lb/hr

$\rho_G =$ ________ lb/ft^3 (refer to Appendix B.9, reference 8, for calculating this variable)

$\rho_L =$ ________ lb/ft^3 (from reference 18)

$ABS = (L/G)(\rho_G/\rho_L)^{0.5}$

$ABS =$ ________

b. From Figure 4-17, determine the value of the ordinate (ORD) at flooding conditions.

$ORD =$ ________

c. For the type of packing used, determine the packing constants from reference 21:

$a =$ ________

$\epsilon =$ ________

Determine μ_L (from reference 18):

$\mu_L =$ ________ cp

d. Use Equation 4.7-8 to calculate $G_{area,f}$:

$$G_{area,f} = \{[ORD\ \rho_G \rho_L\ g_c]/[(a/\epsilon^3)(\mu_L)^{0.2}]\}^{0.5}$$

$G_{area,f}$ = ________ lb/sec-ft²

e. Assume a value for the fraction of flooding velocity for the proposed design:

f = ________

Use Equation 4.7-9 to calculate G_{area}:

$$G_{area} = f\ G_{area,f}$$

G_{area} = ________ lb/hr-ft²

f. Use Equation 4.7-10 to calculate the column cross-sectional area:

$$A_{column} = G/(3{,}600\ G_{area})$$

A_{column} = ________ ft²

g. Use Equation 4.7-11 to calculate the column diameter:

$$D_{column} = 1.13\ (A_{column})^{0.5}$$

D_{column} = ________ ft

4.7.3.3 Column Height

a. Use Equation 4.7-13 or Figure 4-18 to calculate N_{OG}:

Using Equation 4.7-13:

HAP_e = ________ ppmv

$$HAP_o = HAP_e\ (1 - 0.01\ RE)$$

HAP_o = ________ ppmv

$$N_{OG} = \ln\{(HAP_e/HAP_o)[1 - (1/AF)] + (1/AF)\}/[1 - (1/AF)]$$

N_{OG} = ________

Using Figure 4-18:

HAP_e/HAP_o = ________

At HAP_e/HAP_o and $1/AF = 1/1.6 = 0.63$, determine N_{OG}:

N_{OG} = ________

b. Use Equations 4.7-14, -15, and -16 to calculate H_G, H_L, and H_{OG}. Determine the packing constants in Equation 4.7-15 using Tables B.9-2 and -3, reference 8, or reference 12:

b = ________ c = ________ d = ________

Y = ________ s = ________

Determine Sc_G and Sc_L using Tables B.9-4 and -5, reference 8, or reference 12:

Sc_G = ________

Sc_L = ________

$L'' = L/A_{column}$

L'' = ________ lb/hr-ft²

μ_L'' = ________ lb/hr-ft (from reference 18)

Calculate H_G and H_L:

$$H_G = [b\,(3{,}600\ G_{area})^c/(L'')^d]\,(Sc_G)^{0.5}$$

H_G = ________ ft

$$H_L = Y(L''/\mu_L'')^s\,(Sc_L)^{0.5}$$

H_L = ________ ft

Calculate H_{OG} using $AF = 1.6$:

$$H_{OG} = H_G + (1/AF)\,H_L$$

H_{OG} = ________ ft

c. Use Equation 4.7-12 to calculate Ht_{column}:

$$Ht_{column} = N_{OG}\,H_{OG}$$

Ht_{column} = ________ ft

d. Use Equation 4.7-18 to calculate Ht_{total}:

$$Ht_{total} = Ht_{column} + 2 + (0.25\ D_{column})$$

Ht_{total} = __________ ft

e. Use Equation 4.7-19 to calculate Wt_{column}:

$$Wt_{column} = (48\ D_{column} \times Ht_{total}) + 39(D_{column})^2$$

Wt_{column} = __________ lb

f. Use Equation 4.7-20 to calculate $V_{packing}$:

$$V_{packing} = 0.785(D_{column})^2 \times Ht_{column}$$

$V_{packing}$ = __________ ft^3

4.7.3.4 Pressure Drop Through the Column

a. Use Equation 4.7-21 to calculate ΔP_a:

Determine the constants using Table B.9-6, reference 8, or reference 12:

g = __________

r = __________

$$\Delta P_a = g \times 10^{-8}\ [10^{(rL''/\rho_L)}](3{,}600\ G_{area})^2/\rho_G$$

ΔP_a = __________ lb/ft^2-ft

b. Use Equation 4.7-22 to calculate ΔP_{total}:

$$\Delta P_{total} = \Delta P \times Ht_{column}$$

ΔP_{total} = __________ lb/ft^2

ΔP_{total} (1/5.2) = __________ in H_2O

4.7.4 Evaluation of Permit Application

Compare the results from the calculations and the values supplied by the permit applicant using Table 4-9. If the calculated values are different from the reported values, the differences may be due to the assumptions involved in the calculations. Therefore, discuss the details of the proposed design with the applicant.

If the calculated values agree with the reported values, then the design of the proposed absorber system may be considered appropriate based on the assumptions made in this handbook.

Table 4-9. Comparison of Calculated Values and Values Supplied by the Permit Applicant for Absorption

	Calculated Value	Reported Value
Solvent flow rate, L_{gal}	...	...
Column diameter, D_{column}	...	...
Column height, Ht_{column}	...	...
Total column height, Ht_{total}	...	...
Packing volume, $V_{packing}$	...	...
Pressure drop, ΔP_{total}	...	...
Column weight, Wt_{column}	...	...

Appendix C.8
Calculation Sheet for Condensation

4.8.1 Data Required

HAP emission stream characteristics:

1. Maximum flow rate, Q_e = ________ scfm
2. Temperature, T_e = ________ °F
3. HAP = ________
4. HAP concentration, HAP_e = ________ ppmv
5. Moisture content, M_e = ________ %
6. Pressure, P_e = ________ mm Hg

Required removal efficiency, RE = ________ %

In the case of a permit review for a condenser, the following data should be supplied by the applicant:

Condenser system variables at standard conditions (70°F, 1 atm):

1. Reported removal efficiency, $RE_{reported}$ = ________ %
2. Emission stream flow rate, Q_e = ________ scfm
3. Temperature of emission stream, T_e = ________ °F
4. HAP = ________
5. HAP concentration, HAP_e = ________ ppmv
6. Moisture content, M_e = ________ %
7. Temperature of condensation, T_{con} = ________ °F
8. Coolant used = ________
9. Temperature of inlet coolant, $T_{cool,i}$ = ________ °F
10. Coolant flow rate, $Q_{coolant}$ = ________ lb/hr
11. Refrigeration capacity, Ref = ________ tons
12. Condenser surface area, A_{con} = ________ ft^2

4.8.2 Pretreatment of the Emission Stream

Check to see if moisture content of the emission stream is high. If it is high, dehumidification is necessary. This can be carried out in a heat exchanger prior to the condenser.

4.8.3 Condenser System Design Variables

The key design variable is the condensation temperature. Coolant selection will be based on this temperature.

In evaluating a permit application, use Table 4-10 to determine if the applicant's values for T_{con}, coolant type, and $T_{cool,i}$ are appropriate:

T_{con} = ________ °F

Coolant type = ________

$T_{cool,i}$ = ________ °F

If they are appropriate, proceed with the calculations. Otherwise, reject the proposed design. The reviewer may then wish to follow the calculation procedure outlined below.

4.8.4 Determination of Condenser System Design Variables

4.8.4.1 Estimation of Condensation Temperature

a. Use Equation 4.8-1 to calculate $P_{partial}$:

$$P_{partial} = 760\{(1 - 0.01\ RE)/[1 - (RE \times 10^{-8}\ HAP_e)]\}HAP_e \times 10^{-6}$$

$P_{partial}$ = ________ mm Hg

b. Use Figure 4-20 to determine T_{con}:

T_{con} = ________ °F

4.8.4.2 Selection of Coolant

Use Table 4-10 to specify the coolant (also see references 18 and 27):

Coolant = ________

4.8.4.3 Condenser Heat Load

1. a. Use Equation 4.8-2 to calculate $HAP_{e,m}$:

$$HAP_{e,m} = (Q_e/387)\ HAP_e \times 10^{-6}$$

$HAP_{e,m}$ = ________ lb-moles/min

b. Use Equation 4.8-3 to calculate $HAP_{o,m}$:

$$HAP_{o,m} = (Q_e/387)[1-(HAP_e \times 10^{-6})][P_{vapor}/(P_e - P_{vapor})]$$

where $P_{vapor} = P_{partial}$

$HAP_{o,m}$ = ________ lb-moles/min

c. Use Equation 4.8-4 to calculate HAP_{con}:

$$HAP_{con} = HAP_{e,m} - HAP_{o,m}$$

HAP_{con} = ________ lb-moles/min

2. a. Calculate heat of vaporization (ΔH) of the HAP from the slope of the graph $[\ln(P_{vapor})]$ vs $[1/(T_{con} + 460)]$ for the P_{vapor} and T_{con} ranges of interest. See Appendix B.10, reference 8, for details.

ΔH = ________ Btu/lb-mole

b. Use Equation 4.8-5 to calculate H_{con}:

$$H_{con} = HAP_{con}[\Delta H + \bar{C}p_{HAP}\,(T_e - T_{con})]$$

where $\bar{C}p_{HAP}$ can be obtained from references 18 and 27.

H_{con} = ________ Btu/min

c. Use Equation 4.8-6 to calculate H_{uncon}:

$$H_{uncon} = HAP_{o,m}\,\bar{C}p_{HAP}(T_e - T_{con})$$

H_{uncon} = ________ Btu/min

d. Use Equation 4.8-7 to calculate H_{noncon}:

$$H_{noncon} = [(Q_e/387) - HAP_{e,m}]\,\bar{C}p_{air}\,(T_e - T_{con})$$

where $\bar{C}p_{air}$ can be obtained from Table B.4-1, reference 8.

H_{noncon} = ________ Btu/min

3. a. Use Equation 4.8-8 to calculate H_{load}:

$$H_{load} = 1.1 \times 60\,(H_{con} + H_{uncon} + H_{noncon})$$

H_{load} = ________ Btu/hr

4.8.4.4 Condenser Size

Use Equation 4.8-9 to calculate A_{con}:

$A_{con} = H_{load}/U\ \Delta T_{LM}$

where ΔT_{LM} is calculated as follows:

$$\Delta T_{LM} = [(T_e - T_{cool,o}) - (T_{con} - T_{cool,i})]/\ln[(T_e - T_{cool,o})/(T_{con} - T_{cool,i})]$$

Assume: $T_{cool,i} = T_{con}$-15, and $T_{cool,o} - T_{cool,i} = 25°F$

$T_{cool,i}$ = ____________ °F

$T_{cool,o}$ = __________ °F

ΔT_{LM} = __________ °F

Assume: U = 20 Btu/hr-ft^2 -°F (if no other estimate is available).

A_{con} = __________ ft^2

4.8.4.5 Coolant Flow Rate

Use Equation 4.8-10 to calculate $Q_{coolant}$:

$$Q_{coolant} = H_{load}/[\underline{C}p_{coolant}\ (T_{cool,o} - T_{cool,i})]$$

The value for $\underline{C}p_{coolant}$ for different coolants can be obtained from references 18 or 27. If water is used as the coolant, $\underline{C}p_{water}$ can be taken as 1 Btu/lb-°F.

$\underline{C}p_{coolant}$ = __________ Btu/lb-°F

$Q_{coolant}$ = __________ lb/hr

4.8.4.6 Refrigeration Capacity

Use Equation 4.8-11 to calculate Ref:

$$Ref = H_{load}/12{,}000$$

Ref = __________ tons

4.8.4.7 Recovered Product

Use Equation 4.8-12 to calculate Q_{rec}:

$$Q_{rec} = 60 \times HAP_{con} \times MW_{HAP}$$

Q_{rec} = __________ lb/hr

4.8.5 Evaluation of Permit Application

Compare the results from the calculations and the values supplied by the permit applicant using Table 4-11. If the calculated values T_{con}, coolant type, $Q_{coolant}$, A_{con}, Ref, and

Q_{rec} are different from the reported values for these variables, the differences may be due to the assumptions involved in the calculations. Discuss the details of the proposed design with the permit applicant.

If the calculated values agree with the reported values, then the design and operation of the proposed condenser system may be considered appropriate based on the assumptions made in this handbook.

Table 4-11. Comparison of Calculated Values and Values Supplied by the Permit Applicant for Condensation

	Calculated Value	Reported Value
Condensation temperature, T_{con}	. . .	. . .
Coolant type	. . .	. . .
Coolant flow rate, $Q_{coolant}$	. . .	. . .
Condenser surface area, A_{con}	. . .	. . .
Refrigeration capacity, Ref	. . .	. . .
Recovered product, Q_{rec}	. . .	. . .

Appendix C.9
Calculation Sheet for Fabric Filters

4.9.1 Data Required

HAP emission stream characteristics:

1. Flow rate, $Q_{e,a}$ = __________ acfm
2. Moisture content, M_e = __________ % (vol)
3. Temperature, T_e = __________ °F
4. Particle mean diameter, Dp = __________ μm
5. SO_3 content = __________ ppm (vol)
6. Particulate content = __________ grains/scf
7. HAP content = __________ % (mass)

In the case of a permit review, the following data should be supplied by the applicant:

1. Filter fabric material __________
2. Cleaning method (mechanical shaking, reverse air, pulse-jet) __________
3. Air-to-cloth ratio __________ ft/min
4. Baghouse construction configuration (open pressure, closed pressure, closed suction) __________

4.9.2 Pretreatment Considerations

If emission stream temperature is not from 50° to 100°F above the dew point, pretreatment is necessary (see Section 3.3.1 and Appendix B.1). Pretreatment will cause two of the pertinent emission stream characteristics to change; list the new values below.

1. Maximum flow rate at actual conditions, $Q_{e,a}$ = __________ acfm
2. Temperature, T_e = __________ °F

4.9.3 Fabric Filter System Design Variables

1. Fabric Type(s) (use Table 4-12):

   a. __________

   b. __________

   c. __________

2. Cleaning Method(s) (Section 4.8.3.2):

a. ______________________

b. ______________________

3. Air-to-cloth ratio, point or range (Table 4-14) __________ ft/min

4. Net cloth area, A_{nc}:

$$A_{nc} = Q_{e,a} / (A/C \text{ ratio})$$

where:

A_{nc} = net cloth area, ft^2

$Q_{e,a}$ = maximum flow rate at actual conditions, acfm

A/C ratio = air-to-cloth ratio, ft/min

A_{nc} = __________ / __________

A_{nc} = __________ ft^2

5. Gross cloth area, A_{tc}:

$$A_{tc} = A_{nc} \times \text{Factor}$$

where:

A_{tc} = gross cloth area, ft^2

Factor = value from Table 4-15, dimensionless

A_{tc} = __________ x __________

A_{tc} = __________ ft^2

6. Baghouse configuration ______________________

4.9.4 Evaluation of Permit Application

Using Table 4-16, compare the results from this section and the data supplied by the permit applicant. As pointed out in the discussion on fabric filter design considerations, the basic design parameters are generally selected without the involved, analytical approach that characterizes many other control systems, such as an absorber system (Section 4.7). Therefore, in evaluating the reasonableness of any system specifications on a permit application, the reviewer's main task will be to examine each parameter in terms of its compatibility with the gas stream and particulate conditions and with the other selected parameters. The following questions should be asked:

1. Is the temperature of the emission stream entering the baghouse within 50° to 100°F above the stream dew point?

2. Is the selected fabric material compatible with the conditions of the emission stream; that is, temperature and composition (see Table 4-12)?

3. Is the baghouse cleaning method compatible with the selected fabric material and its construction; that is, material type and woven or felted construction (see Section 4.9.3.2 and Table 4-13)?

4. Will the selected cleaning mechanism provide the desired control?

5. Is the A/C ratio appropriate for the application; that is, type of dust and cleaning method used (see Table 4-14)?

6. Are the values provided for the gas flow rate, A/C ratio, and net cloth area consistent? The values can be checked with the following equation:

$$\text{A/C ratio} = \frac{Q_{e,a}}{A_{nc}}$$

where:

A/C ratio = air-to-cloth ratio, ft/min

$Q_{e,a}$ = emission stream flow rate at actual conditions, acfm

A_{nc} = net cloth area, ft^2

7. Is the baghouse configuration appropriate; that is, is it a negative-pressure baghouse?

Appendix C.10
Calculation Sheet for Electrostatic Precipitators

4.10.1 Data Required

HAP emission stream characteristics:

1. Flow rate, $Q_{e,a}$ = ________ acfm
2. Emission stream temperature, T_e = ________ °F
3. Particulate content = ________ grams/scf
4. Moisture content, M_e = ________ % (vol)
5. HAP content = ________ % (mass)
6. Drift velocity of particles, U_d = ________ ft/s
7. Collection efficiency, CE = ________ % mass

In case of a permit review, the following data should be supplied by the applicant. The design considerations in this section will then be used to check the applicant's design.

1. Reported collection efficiency = ________ %
2. Reported drift velocity of particles = ________ ft/sec
3. Reported collection plate area = ________ ft^2

4.10.2 Pretreatment of Emission Stream

If the emission stream temperature is not from 50° to 100°F above the dew point, pretreatment is necessary (see Section 3.3.1 and Appendix B.1). Pretreatment will cause two of the pertinent emission stream characteristics to change; list the new values below.

1. Maximum flow rate at actual conditions, $Q_{e,a}$ = ________ acfm
2. Temperature, T_e = ________ °F

4.10.3 ESP Design Variables

Collection plate area is a function of the emission stream flow rate, drift velocity of the particles (Table 4-17), and desired control efficiency. The variables are related by the Deutsch-Anderson equation:

$$A_p = \frac{-Q_{e,a}}{60 \times U_d} \times \ln(1 - CE)$$

where:

A_p = collection plate area, ft^2

$Q_{e,a}$ = emission stream flow rate at actual conditions as it enters the control device, acfm

U_d = drift velocity of particles, ft/sec

CE = required collection efficiency, decimal fraction

$$A_p = \frac{(-\_\_\_\_\_\_)}{60 \times (\_.\_\_\_\_)} \times \ln(1 - 0.\_\_\_\_)$$

A_p = __________ ft^2

4.10.4 Evaluation of Permit Application

Using Table 4-18, compare the results from this section and the data supplied by the permit applicant. In evaluating the reasonableness of ESP design specifications in a permit application, the main task will be to examine each parameter in terms of its capability with the gas stream conditions.

If the applicant's collection plate area is less than the calculated area, the discrepancy will most likely be the selected drift velocity. Further discussions with the permit applicant are recommended to evaluate the design assumptions and to reconcile any apparent discrepancies.

Appendix C.11
Calculation Sheet for Venturi Scrubbers

4.11.1 Data Required

HAP emission stream characteristics:

1. Flow rate $Q_{e,a}$ = ________ acfm
2. Temperature, T_e = ________ °F
3. Moisture content, M_e = ________ %
4. Required collection efficiency, CE = ________ %
5. Particle mean diameter, D_p = ________ μm
6. Particulate content = ________ grains/scf
7. HAP content = ________ % (mass)

In the case of a permit review, the following data should be supplied by the applicant:

1. Reported pressure drop across venturi = ________ in H_2O
2. An applicable performance curve for the venturi scrubber
3. Reported collection efficiency = ________ %

4.11.2 Pretreatment of Emission Stream

If the emission stream temperature is not from 50° to 100°F above the dew point, pretreatment is necessary (see Section 3.3.1 and Appendix B.1). Pretreatment will cause two of the pertinent emission stream characteristics to change; list the new values below:

1. Maximum flow rate at actual conditions, $Q_{e,a}$ = ________ acfm
2. Temperature, T_e = ________ °F

4.11.3 Venturi Scrubber Design Variables

4.11.3.1 Venturi Scrubber Pressure Drop

The pressure drop across the venturi (ΔP_v) can be estimated through the use of a venturi scrubber performance curve (Figure 4-22) and known values for the required collection efficiency (CE) and the particle mean diameter (D_p).

ΔP_v = ________ in H_2O

If the estimated ΔP_v is greater than 80 in H_2O, assume that the venturi scrubber cannot achieve the desired control efficiency.

4.11.3.2 Materials of Construction

Select the proper material of construction by contacting a vendor, or as a lesser alternative, by using Table 4-20.

Material of construction ____________________

4.11.4 Sizing of Venturi Scrubbers

Some performance curves and cost curves are based on the saturated gas flow rate ($Q_{e,s}$). If $Q_{e,s}$ is needed, it can be calculated as follows:

$$Q_{e,s} = Q_{e,a} \times (T_{e,s} + 460)/(T_e + 460)$$

where:

$Q_{e,s}$ = saturated emission stream flow rate, acfm

$T_{e,s}$ = temperature of the saturated emission stream, °F

Use Figure 4-22 to determine $T_{e,s}$; the moisture content of the emission stream (M_e) must be in units of lbs H_2O/lbs dry air.

Convert M_e (% vol) to units of lbs H_2O/lbs dry air, decimal fraction:

$(M_e/100) \times (18/29)$ = ________ lb H_2O/lb dry air

From Figure 4-22:

$T_{e,s}$ = ________ °F

$Q_{e,s}$ = (__________) x (__________ + 460) / (__________ + 460)

$Q_{e,s}$ = ________ acfm

4.11.5 Evaluation of Permit Application

Using Table 4-21, compare the results of this section and the data supplied by the permit applicant. Compare the estimated ΔPv and the reported pressure drop across the venturi, as supplied by the permit applicant.

If the estimated and reported values differ, the differences may be due to the applicant's use of another performance chart, or a discrepancy between the required and reported collection efficiencies. Discuss the details of the design and operation of the system with the applicant. If there are no differences between the estimated and reported values for ΔPv, the design and operation of the system can be considered appropriate based on the assumptions employed in this handbook.

Table 4-21. Comparison of Calculated Values and Values Supplied by the Permit Applicant for Venturi Scrubbers

	Calculated Value	Reported Value
Particle Mean Diameter, D_p	. . .	. . .
Collection efficiency, CE	. . .	. . .
Pressure drop across venturi, ΔP_v	. . .	. . .

Appendix C.12
Capital and Annualized Cost Calculation Worksheet

Table C.12-1. Preliminary Calculations for Capital Cost Algorithm

(1) Calculation of Duct Diameter, D_{duct} (in)

$$D_{duct} = 12 \left(\frac{4}{\pi} \times \frac{Q_{e,a}}{U_{duct}} \right)^{\frac{1}{2}}$$

where:

$Q_{e,a}$ = emission stream flow rate at actual conditions, acfm

U_{duct} = velocity of gas stream in duct, ft/min

$$D_{duct} = 12 \left(\frac{4}{\pi} \times \frac{(\qquad)}{(\qquad)} \right)^{\frac{1}{2}} = \underline{\qquad\qquad} \text{ in}$$

If velocity of gas stream in duct is unknown, use 2,000 ft/min; the equation then becomes:

$$D_{duct} = 0.3028\,(Q_{e,a})^{\frac{1}{2}}$$

$$D_{duct} = 0.3028\,(\underline{\qquad\qquad})^{\frac{1}{2}} = \underline{\qquad\qquad} \text{ in}$$

(2) Calculation of Stack Diameter, D_{stack} (in)

$$D_{stack} = 12 \left(\frac{4}{\pi} \times \frac{Q_{fg,a}}{U_{stack}} \right)^{\frac{1}{2}}$$

where:

$Q_{fg,a}$ = actual flue gas flow rate, acfm

U_{stack} = velocity of gas in stack, ft/min

$$D_{stack} = 12 \left(\frac{4}{\pi} \times \frac{(\qquad)}{(\qquad)} \right)^{\frac{1}{2}} = \underline{\qquad\qquad} \text{ in}$$

The gas stream velocity in the stack should be at least 4,000 ft/min. If velocity is unknown, use 4,000 ft/min; the equation then becomes:

$$D_{stack} = 0.2141\,(Q_{fg,a})^{\frac{1}{2}}$$

$$D_{stack} = 0.2141\,(\underline{\qquad\qquad})^{\frac{1}{2}} = \underline{\qquad\qquad} \text{ in}$$

(3) Calculation of Total System Pressure Drop, ΔPt (in H_2O)

$$\Delta P_t = \Delta P_{duct} + \Delta P_{stack} + \Delta P_{device\ \#1} + \Delta P_{device\ \#2} + \Delta P_{device\ \#3}$$

(Note: See Table 5-7 for ΔP values.)

$$\Delta Pt = \underline{\qquad\qquad} + \underline{\qquad\qquad} + \underline{\qquad\qquad} + \underline{\qquad\qquad} + \underline{\qquad\qquad} = \underline{\qquad\qquad} \text{ in } H_2O$$

Table C.12-2. Estimate of Capital Costs in Current Dollars

Cost Elements	Figure or Table Cost	Escalation Factor (Current FE/Base FE) see Table 5-2	Current Cost
1. Major Equipment Purchase Cost			
Thermal Incinerator[a]	$ ______	× (______/______) =	$ ______
Heat Exchanger[b]	$ ______	× (______/______) =	$ ______
Catalytic Incinerator[c]	$ ______	× (______/______) =	$ ______
Catalyst[c], V_{cat} = ______	ft³ x ______ $/ft³	× (______/______) =	$ ______
Carbon Adsorber[d]	$ ______	× (______/______) =	$ ______
Carbon[d], C_{req} = ______	lb x ______ $/lb	× (______/______) =	$ ______
Absorber[e]	$ ______	× (______/______) =	$ ______
Platforms and Ladders[e]	$ ______	× (______/______) =	$ ______
Packing[e], V_{pack} = ______	ft³ x ______ $/ft³	× (______/______) =	$ ______
Condenser[f]	$ ______	× (______/______) =	$ ______
Refrigerant[f]	$ ______	× (______/______) =	$ ______
Fabric Filter[g]	$ ______	× (______/______) =	$ ______
Bags[g], A_{tc} = ______	ft² x ______ $/ft²	× (______/______) =	$ ______
ESP[h]	$ ______	× (______/______) =	$ ______
Venturi Scrubber[i]	$ ______	× (______/______) =	$ ______
Design Factors[i]	______ (Thickness Factor)	× ______ (Composition Factor) =	$ ______
		SUBTOTAL	$ ______
2. Auxiliary Equipment Purchase Cost			
Ductwork[j]	$______ × ______ (Length)	× (______/______) –	$ ______
Fan[k]	$ ______	× (______/______) =	$ ______
Motor[l]	$ ______ (Fan Current Cost)	× 0.15 =	$ ______
Stack[m]	$ ______	× (______/______) =	$ ______
		SUBTOTAL	$ ______
3. Pre-Total Purchase Equipment Cost	Item 1 Subtotal + Item 2 Subtotal		$ ______
Adjustments[n]	(Item 3) x -0.091		$ ______
4. TOTAL Purchase Equipment Cost	Item 3 + Adjustments		$ ______
5. Instrumentation and Controls	10% of Item 4		$ ______
6. Freight and Taxes	8% of Item 4		$ ______
7. TOTAL Purchased Cost	Item 4 + Item 5 + Item 6		$ ______
8. TOTAL CAPITAL COSTS	F[o] x (Item 7); where F = ______		$ ______

Footnotes to Table C.12-2

[a] Thermal Incinerator: Figure 5-1, includes fan plus instrumentation and control costs for thermal incinerators, in addition to the major equipment purchased cost. Additional auxiliary equipment (ductwork and stack) purchased costs and costs of freight and taxes must be added to obtain the total purchased cost.

[b] Heat Exchangers: If the HAP control system requires a heat exchanger, obtain the cost from Figure 5-2, escalate this cost using the appropriate factor, and add to the major equipment purchased cost.

[c] Catalytic Incinerator: Figure 5-3 provides the cost of a catalytic incinerator, less catalyst costs. The "Table" catalyst cost is estimated by multiplying the volume of catalyst required (V_{cat}) by the catalyst cost factor ($/ft^3) found on Table 5-3. Catalyst costs, all auxiliary equipment (ductwork, fan, and stack) purchased costs and the cost of instrumentation and controls, and freight and taxes must be added to obtain the total purchased cost.

[d] Carbon adsorber: Figure 5-4 (packaged carbon adsorber systems) includes the cost of carbon, beds, fan and motor, instrumentation and controls, and a steam regenerator. Additional auxiliary equipment (ductwork and stack) purchased costs and costs of freight and taxes must be added to obtain the total purchased cost. Figure 5-5 (custom carbon adsorber systems) includes beds, instrumentation and controls, and a steam regenerator, less carbon. The "Table" carbon cost for custom carbon adsorbers is estimated by multiplying the weight of carbon required (C_{req}) by the carbon cost factor ($/lb) found on Table 5-3. Costs of carbon, all auxiliary equipment (duct, fan, stack) purchased costs, and freight and taxes must be added to obtain the total purchased cost.

[e] Absorber: Figure 5-6 does not include the cost of packing, platforms, and ladders. The cost of platforms and ladders (Figure 5-7) and packing must be added to obtain the major purchased equipment cost. The "Table" packing cost is estimated by multiplying the volume of packing required (V_{pack}) by the appropriate packing cost factor found on Table 5-4. All auxiliary equipment (ductwork, fan, and stack) purchased costs, and costs of freight and taxes must be added to obtain the total purchased cost.

[f] Condenser Systems: Figure 5-8 yields total capital costs for cold water condenser systems. For systems needing refrigerant, the applicable cost from Figure 5-9 must be added to obtain the total capital costs. In either case, the escalated cost estimate is then placed on Line 8, "TOTAL CAPITAL COSTS."

[g] Fabric Filter Systems: Figure 5-10 gives the cost of a negative pressure, insulated baghouse. The curve does not include bag costs. The "Table" bag cost is estimated by multiplying the gross cloth area required (A_{tc}) by the appropriate bag cost factor found on Table 5-5. Bag costs, all auxiliary equipment (duct, fan, and stack) purchased costs, the cost of instrumentation and controls, and freight and taxes must be added to obtain the total purchased cost.

[h] Electrostatic Precipitators: Figure 5-11 provides the cost for an insulated ESP. All auxiliary equipment (duct, fan, and stack) purchased costs, the cost of instrumentation and controls, and freight and taxes must be added to obtain the total purchased cost.

[i] Venturi Scrubber: Figure 5-12 includes the cost of instrumentation and controls in addition to the major equipment purchased cost. This cost curve is based on a venturi scrubber constructed from 1/8-inch carbon steel. Figure 5-13 is used to determine if 1/8-inch steel is appropriate for a given application (use the higher curve). If thicker steel is required, Figure 5-14 yields an adjustment factor for various steel thicknesses; this factor is used to escalate the cost obtained from Figure 5-12. In addition, if stainless steel is required (see Section 4.11.3.2) multiply the scrubber cost estimate by 2.3 for 304L stainless steel or by 3.2 for 316L stainless steel. Costs of all auxiliary equipment (ductwork, fan, and stack) and freight and taxes must be added to obtain the total purchased cost.

[j] Ductwork: Figure 5-15 gives the cost of straight ductwork made of carbon steel for various thicknesses, based on the required duct diameter. Figure 5-16 gives the cost of straight ductwork made of stainless steel for various thicknesses, based on the required duct diameter. Preliminary calculations (duct diameter, see Table C.12-1) are necessary to estimate ductwork costs.

[k] Fan: Figure 5-17 gives the cost of a fan based on the gas flow rate at actual conditions and the HAP control system pressure drop (in inches of H_2O). The applicable fan class is also based on the HAP control system pressure drop. Calculation of the total system pressure drop is presented in Table C.12-1.

[l] The cost of a motor is estimated as 15% of the fan cost.

[m] Stack: Figure 5-18 gives the cost of a carbon steel stack at various stack heights and diameters. Figure 5-19 gives the price of a stainless steel stack at various stack heights and diameters. Preliminary calculations (stack diameter, see Table C.12-1) are necessary to estimate stack costs. For both figures, use the curve that best represents the calculated diameter.

[n] For thermal incinerators, carbon adsorbers, and venturi scrubbers, the purchase cost curve includes the cost for instrumentation and controls. This cost (i.e., the "Adjustment") must be subtracted out to estimate the total purchased equipment cost. This is done by adding the Item 1 subtotal and the Item 2 subtotal and multiplying the result by -0.091. This value is added to the preliminary total purchased equipment cost to obtain the total purchased equipment cost. For all other major equipment, the "Adjustment" equals zero.

[o] Obtain factor "F" from "TOTAL" line in Table 5-8.

Table C.12-3. Preliminary Calculations for Annualized Cost Algorithm

(1) Calculation of Annual Electricity Requirement, AER (Line 5, Table C.12-6)

a. Fan Electricity Requirement, FER

$FER = 0.0002\ (Q_{fg,a}) \times \Delta P \times HRS$

where:
$Q_{fg,a}$ = actual flue gas flow rate, acfm

ΔP = total HAP control system pressure drop, in H_2O (see Table 5-7)

HRS = annual operating hours, hr
(Note: Use 8,600 unless otherwise specified.)

FER = 0.0002 (________) x ______ x ______ = ________ kWh

b. Baghouse Electricity Requirement, BER

(Note: Assume 0.0002 kW are required per ft^2 of gross cloth area.)

$BER = 0.0002\ (A_{tc}) \times HRS$

where:
A_{tc} = gross cloth area required, ft^2

BER = 0.0002 (________) x ________ = ________ kWh

c. ESP Electricity Requirement, EER

(Note: Assume 0.0015 kW are required per ft^2 of collection area.)

$EER = 0.0015\ (A_p) \times HRS$

where:
A_p = collection plate area, ft^2

EER = 0.0015 (________) x ________ = ________ kWh

d. Annual Electricity Requirement, AER

$AER = FER + BER + EER$

AER = ________ + ________ + ________ = ________ kWh

(2) Calculation of Capital Recovery Factor, CRF (Line 18, Table C.12-6)

$CRF = [i(l + i)^n] / [(l + i)^n - 1]$

where:
i = interest rate on borrowed capital, decimal fraction
(Note: Unless otherwise specified use 10%.)

n = control device lifetime, years (see Table 5-12)

CRF = [______ x (1 + ______) (   )] / [(1 + ______)(   )-1] = ______

(3) Calculation of Annual Operator Labor, OL (Line 9, Table C.12-6)

OL = (HRS) (operator hours per shift) / (operating hours per shift)
(Note: Obtain operator hr/shift value from Table 5-12)

OL = (______) x (______) / (______) = ______ hr

(4) Calculation of Annual Maintenance Labor, ML (Line 11, Table C.12-6)

ML = (HRS) (maintenance hours per shift) / (operating hours per shift)
(Note: Obtain maintenance hr/shift value from Table 5-12)

ML = (______) x (______) / (______) = hr

Table C.12-4. Additional Utility Requirements

(1) Fuel Requirement for Incinerators (Line 1 or Line 2, Table C.12-6)

(Note: The design sections for thermal and catalytic incinerators are developed under the assumption that natural gas is used as the supplementary fuel. Fuel oil could be used, however, the use of natural gas is normal industry practice. If fuel oil is used, the equation below can be used by replacing Q_f with the fuel oil flow rate in units of gallons per minute. The resultant product of the equation [gallons of fuel oil required] is then used on Line 2 of Table C.12-6.)

Fuel Requirement = 60 (Q_f) x HRS

where:
Q_f = supplementary fuel required, scfm

HRS = annual operating hours, hr
(Note: Use 8,600 hours unless otherwise specified.)

Fuel Requirement = 60 (________) x ________ = ________ ft^3

(2) Steam Requirement for Carbon Adsorber (Line 4, Table C.12-6)

(Note: Assume 4 lb of steam required for each lb of recovered product.)

Steam Requirement = 4 (Q_{rec}) x HRS

where:
Q_{rec} = quantity of HAP recovered, lb/hr

HRS = annual operating hours, hr
(Note: Use 8,600 hours unless otherwise specified.)

Steam Requirement = 4 (________) x ________ = ________ lb

(3) Cooling Water Requirement for Carbon Adsorber (Line 3, Table C.12-6)

(Note: Assume 12 gal of cooling water required per 100 lbs steam.)

Water Requirement = 0.48 (Q_{rec}) x HRS

where:
Q_{rec} = quantity of HAP recovered, lb/hr

HRS = annual operating hours, hr
(Note: Use 8,600 hours unless otherwise specified.)

Water Requirement = 0.48 (________) x ________ = ________ gal

(4) Absorbent Requirement for Absorbers (Line 3 or 6, Table C.12-6)

(Note: Assume no recycle of absorbing fluid [water or solvent].)

Absorbent Requirement = $60 (L_{gal}) \times HRS$

where:
L_{gal} = absorbing fluid flow rate, gal/min

HRS = annual operating hours, hr
(Note: Use 8,600 hours unless otherwise specified.)

Absorbent Requirement = 60 (________) x ________ = ________ gal

(5) Water Requirement for Venturi Scrubbers (Line 3, Table C.12-6)

(Note: Assume 0.01 gal of water required per acf of emission stream.)

Water Requirement = $0.6 (Q_{e,a}) \times HRS$

where:
$Q_{e,a}$ = emission stream flow rate into scrubber, acfm

HRS = annual operating hours, hr
(Note: Use 8,600 hours unless otherwise specified.)

Water Requirement = 0.6 (________) x ________ = ________ gal

Table C.12-5. Estimation of Replacement Parts Annualized Costs

(1) Annualized Catalyst Replacement Costs (Line 7, Table C.12-6)

Over the lifetime of a catalytic incinerator, the catalyst is depleted and must be replaced (assume catalyst lifetime is 3 years):

Annual Catalyst Cost = (Catalyst Current Cost[a]) / 3

Annual Catalyst Cost = (__________) / 3 = $__________

(2) Annualized Carbon Replacement Costs (Line 7, Table C.12-6)

Over the lifetime of a carbon adsorber, the carbon is depleted and must be replaced (assume carbon lifetime is 5 years):

Annual Carbon Cost = (Carbon Current Cost[a]) / 5

Annual Carbon Cost = (__________) / 5 = $__________

(3) Annualized Refrigerant Replacement Costs

Refrigerant in a condenser needs to be replaced periodically due to system leaks; however, the loss rate is typically very low. Therefore, assume the cost of refrigerant replacement is negligible.

(4) Annualized Bag Replacement Costs (Line 7, Table C.12-6)

Over the lifetime of a fabric filter system the bags become worn and must be replaced (assume bag lifetime is 2 years):

Annual Bag Cost = (Bag Current Cost[a]) / 2

Annual Bag Cost = (__________) / 2 = $__________

[a]See Table C.12-2.

Table C.12-6. Estimate of Annualized Costs in Current Dollars

Cost Elements	Units Costs/Factor		Annual Expenditure		Current Dollars
Direct Operating Costs					
1. Natural Gas[a]	$0.00425 per ft^3	x	________ ft^3	=	$________
2. Fuel Oil[a]	$1.025 per gal	x	________ gal	=	$________
3. Water[a]	$0.0003 per gal	x	________ gal	=	$________
4. Steam[a]	$0.00504 per lb	x	________ lb	=	$________
5. Electricity[b]	$0.059 per kWh	x	________ kWh	=	$________
6. Solvent[a]	$________ per gal[c]	x	________ gal	=	$________
7. Replacement Parts	As applicable (see Table C.12-5)				$________
8. Replacement Labor	100% of Line 7				$________
9. Operator Labor[b]	$11.53 per hr	x	________ hr	=	$________
10. Supervision Labor	15% of Line 9				$________
11. Maintenance Labor[b]	$11.53 per hr	x	________ hr	=	$________
12. Maintenance Materials	100% of Line 11				$________
13. SUBTOTAL	Add Items 1 through 12				$________
Indirect Operating Costs					
14. Overhead	80% of Sum of Lines 8, 9, 10, and 11				$________
15. Property Tax	1% of Total Capital Cost[d]				$________
16. Insurance	1% of Total Capital Cost[d]				$________
17. Administration	2% of Total Capital Cost[d]				$________
18. Capital Recovery[b]	(CRF) x Total Capital Cost[d]; where CRF = ________				$________
19. SUBTOTAL	Add Items 14 through 18				$________
20. CREDITS	As applicable (see Section 5.2.3)				$________
NET ANNUALIZED COSTS	Item 13 + Item 19 - Item 20				$________

[a] See Table C.12-4.

[b] See Table C.12-3.

[c] As applicable.

[d] Total Capital Cost from Line 8 of Table C.12-2.